What Others are Saying About *Twelve O'Clock Haiku*

"I have never felt as if I did not 'get' history, but have often worried I did not 'get' poetry. *Twelve O'Clock Haiku* has changed my mind. An excellent collection of not only poems, but also a terrific reference guide, reading list, and life and leadership lessons for all [...] this book should find a home on the shelves and in the offices of those in and out of the military. This book is on target and on time!"

—Brian D. Laslie, command historian, U.S. Air Force Academy, and author of *Architect of Air Power, The Air Force Way of War,* and *Air Power's Lost Cause*

❖ ❖ ❖

"The two languages in which members of the armed forces are most fluent are doctrinal cliché and war movie, and Randy Brown has brilliantly distilled both into key military leadership lessons drawn from the classic WW2 film, 'Twelve O'Clock High.' Alex, the protagonist of 'A Clockwork Orange,' observes 'it's funny how the colours of the like real world only seem really real when you viddy them on the screen.' Perhaps this is why men and women in uniform tend to be such avid consumers of war-themed popular culture—even when they themselves are at war. In *Twelve O'Clock Haiku*, Brown offers a delightful collection of insightful narrative analysis and appropriately metaphysical poems. Haiku might actually be the best poetic form for capturing the semiotics of military life, where so much is so often communicated by so little, because in haiku—as in the military's beloved acronyms, buzzwords, and proverbs—the spark of insight is always found where the words are not. Leadership wisdom, it seems, is only really real when we viddy it in a poem."

—Russell A. Burgos, associate professor, National Defense University

Other Military-themed Books from Middle West Press

anthologies

Our Best War Stories:
Prize-winning Poetry & Prose
from the Col. Darron L. Wright Memorial Awards
Edited by Christopher Lyke

Why We Write:
Craft Essays on Writing War
Edited by Randy Brown
& Steve Leonard

Reporting for Duty:
U.S. Citizen-Soldier Journalism
from the Afghan Surge, 2010-2011
Edited by Randy Brown

poetry collections

Hugging This Rock:
Poems of Earth & Sky, Love & War
by Eric Chandler

Permanent Change of Station and FORCES
by Lisa Stice

Welcome to FOB Haiku:
War Poetry from Inside the Wire,
by Randy Brown, a.k.a. "Charlie Sherpa"

Twelve O'Clock Haiku

Leadership Lessons from Old War Movies & New Poems

Randy Brown
Middle West Press LLC
Johnston, Iowa

Twelve O'Clock Haiku
Copyright © 2022 by Randy Brown

All rights reserved. Except for brief quotations in critical articles or reviews,
no part of this book may be reproduced
without prior written permission from the publisher.

Poetry / Leadership & Ethics / Film History / Military Life

Brown, Randy
Twelve O'Clock Haiku:
Leadership Lessons from Old War Movies & New Poems
ISBN (print): 978-1-953665-10-2
ISBN (e-book): 978-1-953665-11-9
Library of Congress Control Number: 2022941658

Middle West Press LLC
P.O. Box 1153
Johnston, Iowa 50131-9420
www.middlewestpress.com

Special thanks to James Burns of Denver, Colorado!
Your patronage helps publish great military-themed writing!
www.aimingcircle.com

In Memory of
Miss Barbara F. Hess (1938-2018)

teacher, mentor, friend

Contents

Introduction vii

Twelve O'Clock Haiku
Twelve O'Clock Haiku 1

On Lessons-Learned & Maximum Efforts
Epigraph 5
"On Lessons-Learned & Maximum Efforts": An Essay in 12 Acts 6
Interlude 12
Act II 13
Interlude 14
Act III 15
Interlude 16
Act IV 17
Interlude 22
Act V 23
Interlude 26
Act VI 27
Interlude 33
Act VII 34
Interlude 36
Act VIII 37
Interlude 38
Act IX 39
Interlude 41
Act X 42
Interlude 43
Act XI 44
Interlude 46
Act XII 47
Epilogue 53

Other Poems, Other Lessons
- 10 haiku about Operation Desert Storm 57
- sound of a left-handed baseball bat, clapping 59
- toward a poetics of lessons-learned 60
- Kintsugi 62
- Better Hooches and Gardens 63
- a quiet professional professes in haiku 64
- Toward an understanding of war and poetry,
  told (mostly) in aphorisms 66
- an Army lessons-learned analyst writes haiku 68
- Humility 69
- contact print 70
- Icarus over Wisconsin 71
- Clausewitzian nature poem 72

Secondary Targets
- Recommended Books, Movies & Poems 75

Bibliography 86

Editorial Style Notes 96

Acknowledgements 97

A Few Words of Thanks 99

About the Writer 101

Did You Enjoy This Book? 102

Introduction

The *Twelve O'Clock Haiku* project started out as a whimsical experiment in minimalist war poetry, and ended up a "maximum effort" in memory, popular media, and military culture.

It is now a set of 12 micro-poems inspired by a favorite old war movie, an interwoven 12-part essay that explores war and how we come to think about it, and a survey of recommended resources—books, movies, and even poems—for those who might wish to further engage in discussion, contemplation, and exploration of these topics.

For good measure, I have also inserted a section of new and selected poems—not all haiku—organized around a theme of "lessons-learned."

Given its esoteric approach, *Twelve O'Clock Haiku* is precisely targeted at a unique and likely very select set of readers. (If you are reading this, *thank you!*) I hope you enjoy:

- Subverting for humorous purposes the rudimentary haiku form as taught in American elementary schools. (Three lines; 5-7-5 syllables; should allude to a moment observed in nature.)
- Using poetry to create opportunities of mutual empathy, understanding, and discussion among people, only some of whom may have served in uniform.
- Watching old war movies.

As I explore in the project's essay on "lessons-learned," whether as a novel, film, or television series, the "Twelve O'Clock High" franchise provides many opportunities to discuss what it means to be a leader or follower at war—morally, physically, and psychologically. Further, its narratives encourage moments of personal and professional reflection:

- What sacrifices are we truly prepared to make?
- How can we best motivate and lead others?
- How can we learn from missteps and successes?

These are always good questions, I think.

And good grist for poetry, too.

— *Randy "Sherpa" Brown*

Twelve O'Clock Haiku

Twelve O'Clock Haiku

1.
In a shop window
a grin of recognition:
"Target for today?"

2.
He's a first-rate guy
who's gonna bust wide-open
thinking of his boys.

3.
Lean on somebody?
I think they're better than that.
If not, we're dead ducks.

4.
We've got to find out
just how much a man can take:
"Maximum Effort."

5.
Take nice kids and fly
until they can't anymore.
And then, fly some more.

6.
Every deadbeat
gets assigned to a plane called
"Leper Colony."

7.
There's only one hope:
daylight precision bombing
will shorten the war.

8.
Those things are coming—
Replacements ... Combat limits ...
Right now, we hang on.

9.
Forget about home.
Once you accept that you're dead,
it won't be so tough.

10.
Stay in formation.
What is not expendable
is your loyalty.

11.
They all looked alike.
All the dead had just one face;
it was very young.

12.
What happened to him?
State of shock. Complete collapse.
He's up there, with them.

On Lessons-Learned & Maximum Efforts

Epigraph

The moon's brightness—
Does it know
Where the bombing will be?

— Santōka Taneda (1882-1940)
translated by John Stevens[1]

---

[1] Taneda, Santōka, and John Stevens. *Mountain Tasting: Zen Haiku*. Weatherhill, 1991. Poem 152, p. 71. This poem was part of a haiku sequence written during the 1937 Japan-China War.

"On Lessons-Learned & Maximum Efforts"
An Essay in 12 Acts

Act I

When I first encountered the 1949 movie "Twelve O'Clock High," it was my job to help prepare my neighbors to go to war in Iraq and Afghanistan.

At the time, I was working toward the end of my 20 years of service in the Iowa Army National Guard. Despite the "one weekend a month, two weeks a year" stereotype of citizen-soldiers, I was then in uniform nearly every day. I won't bore you with mercurial discussions of reserve duty codes—let's just say it was a federally funded temporary full-time tour, without the possibility of overseas deployment.

(During my career as a reservist, including a 7-month overseas deployment in 2003-2004, I accumulated an approximate total of 5 years of active-duty time.)

In the years following the terrorist attacks of Sept. 11, 2001, the state of Iowa regularly sent units to Iraq, Afghanistan, and other hotspots of varying temperatures. In 2008, we were nearing the peak of the Global War on Terror (GWOT). The largest deployment of Iowa troops since World War II was on the horizon—a brigade combat team comprising more than 3,000 of us—and there was plenty of budget to go along with it. In that context, I was put on orders to use my civilian-acquired journalism skills (I was once the editor of national consumer newsstand "how-to" magazines) to write "lessons-learned" reports.

Despite the pervasive "road to war" mentality at work, it was a cream-puff assignment—a grand thought-experiment from someone up higher in the food chain, made possible by an excess of use-it-or-lose-it wartime funding. We were empowered to act creatively, trusted to discreetly document flaws any we found in the system, and otherwise told to keep out of the way.

In fact, the immediate supervisor to whom we'd been attached wasn't quite sure what to do with us. We were issued a little office building, located on an older part of Camp Dodge—far away from the generals' flags at Iowa Joint Forces Headquarters. Our standing orders were minimal: First, that 80 percent of our job was to figure out what our job was supposed to be. Second, that we could invite ourselves to any event we thought of interest. (Evoking the fictional spy James Bond, we called that edict our "License to Kill and/or Attend Meetings.")

My crack two-person multi-media team—one print journalist (me) partnered with one broadcast journalist—documented innovations we observed in the field, as Iowa units prepared for deployments to Afghanistan and Iraq. We then shared those insights with other units, as they prepared for similar missions. We were never told what to write, nor do I recall any instances of censorship. We produced videos, published monthly newsletters, and generated white-paper reports. We disseminated our information products first throughout the state, and also submitted them to the Center for Army Lessons Learned ("CALL") at Fort Leavenworth, Kansas. There, they were shared with an Army-wide network of other "Lessons-Learned Integration (L2I) Analysts" embedded at various schoolhouses, units, and commands.

In many ways, writing lessons-learned paralleled my experiences in civilian journalism. As an editor for *Better Homes and Gardens* specialty publications, I'd produced hundreds of stories about happy homeowners, builders, and architects. Before that, I'd been an editor for equally esoteric national trade magazines, covering commercial and governmental building construction, management, and operations. There, I'd regularly written what were effectively organizational profiles: behind-the-scenes explorations of hotels and hospitals, office complexes and college campuses, theme parks and military bases.

In both my civilian and military writing careers, I now realize, my practice was to metaphorically parachute into an organization, in order to talk with "stakeholders" at all levels. In writing about facilities operations, that meant owners, architects, tenants, and contractors. In writing about military operations, that meant commanders, trainers, trainees ... and more contractors. In each context, the core questions were similar: *What did you set out to do? How did things turn out? What would you do differently, if you did it all again?*

Over lunch one day, an Army officer buddy recommended that I watch the 1949 movie "Twelve O'Clock High." He had previously come across it during one or more "professional military education" (PME) classes required in his career, and he thought it was particularly relevant to our informal conversations about leadership. He was so passionate about the film, he even lent me his personal DVD copy.

It turned out that "Twelve O'Clock High" was my favorite kind of war story: The plot is high-stakes yet human-scaled, and set at a time and place in which victory seems anything but certain. The good guys may not win, and many of them may die trying. Documentary film footage adds gritty realism to the special effects.

"Praise the Lord," I said, "and pass the popcorn."

The movie tells the story of a fledgling U.S. military air component, early in America's direct involvement in World War II Europe. The U.S. Army Air Forces seeks to justify its existence, doctrine, and strategic role. Meanwhile, the force is suffering heavy losses in personnel and equipment. Given my lessons-learned focus—a job in which I imagined myself as a "fly on the wall" observer of organizations under wartime stresses—the film's central questions seemed to echo my own: *What did you set out to do in the war? How did things turn out?*

The 2-hour-and-12-minute film features not only the chisel-jawed actor Gregory Peck as the charismatic Brig. Gen. Frank Savage, but also another cinematic hero of mine, Dean Jagger (pronounced "JAG-ger"). Jagger plays Maj. Harvey Stovall, a citizen-soldier "retread" veteran of World War I, who leverages his civilian-acquired skills as a lawyer in his role as a wickedly creative administrative officer for the fictional 918th Bomb Group. The portrayal earned Jagger a 1950 Academy Award for best-supporting actor.

The character of Frank Savage was inspired by the real-life commander of the U.S. 306th Bomb Group, Col. Frank Armstrong. (The character's surname was meant to be a tribute to Armstrong's Native American heritage.) Eighth Army Air Force commander Maj.

Gen. Ira Eaker twice deployed Armstrong to trouble-shoot hard-luck squadrons that were flying out of England. Armstrong's 6-week assignment as a problem-solving interim commander of the 306th Bomb Group in January 1943 earned him a promotion to brigadier general.[1] He was a fighting general—a flying general.

The fictional Frank Savage was equally inspired by the real-life commander of the 305th Bomb Group, Col. Curtis LeMay. LeMay's innovations in training and tactics—including the designation of lead bombardiers, and arranging planes into staggered "combat box" flying formations to optimize defensive firepower—were quickly implemented throughout the emerging bomber force. He was also known for closed-door debriefings at which bomber crews could say absolutely anything, regardless of rank.[2] After his time in Europe, LeMay eventually would be assigned to command U.S. air forces in the Pacific Theater.

After the war, when "Twelve O'Clock High" was released as a movie, LeMay was the commander of Strategic Air Command ("SAC"), Offutt Air Force Base, Omaha, Nebraska. He later served as U.S. Air Force chief of staff 1961-1965.

Produced by Darryl F. Zanuck and directed by Henry King, the 20th Century Fox film "Twelve O'Clock High" is a certified classic. It was nominated for best picture in the 1950 Academy Awards, and Peck was nominated for best actor. (This was his fourth best-actor nomination. In 1962, he would finally win the title for his portrayal of Atticus Finch in "To Kill a Mockingbird.") In addition to Jagger's win for best-supporting actor, Thomas T. Moulton won an Oscar for best sound recording.

In 1998, "Twelve O'Clock High" was added to the National Film Registry, as established by the U.S. Library of Congress. An archetypical World War II "command drama,"[3] the film has been used as a case study

---

[1] Duffin, Allan T., and Paul Matheis. *The 12 O'Clock High Logbook: The Unofficial History of the Novel, Motion Picture, and TV Series.* Kindle ed. BearManor Media, 2005
[2] Coffey, Thomas M. *Iron Eagle: The Turbulent Life of General Curtis LeMay.* Crown Publishers Inc., 1986. pp. 47-49
[3] Defined by A. Bowdoin Van Riper in "Baa Baa Black Sheep and the Last Stand of the WWII Drama" *American Militarism on the Small Screen,* edited by Anna Froula and Stacy Takacs. (Routledge, 2016): "The [Commanding Officer] is—if not the hero in a

at the Harvard Business School, and reportedly featured in instructional materials in venues such as U.S. Navy's Leadership and Management Training School, the Air Force's College for Enlisted Professional Military Education, and the U.S. Army's Command and General Staff College.[4] It continues to pop up in business magazine articles, essays about national security,[5] and in "Top Ten List" blog posts about classic war movies. In recent popular culture, it was even referenced in the dialogue of 2016's snarky superhero movie "Deadpool."

I eventually had to return my buddy's DVD copy of the movie. And, rather than wear out the disc available at my public library, I soon sought to acquire my own. At a used-book shop, I bought a 2-disc "special edition" DVD published in 2007 for a price around $5. The extra disc contains some documentary material. Earlier DVD editions were published in 2002 and 2005.[6]

I first encountered the TV series "12 O'Clock High" by chance, during some sleepless nights caused by a sore throat. In the feverish dark hours before dawn, I discovered reruns of the 1964-1967 ABC TV series on a syndicated broadcast channel called Heroes and Icons (H&I). It can still be viewed there. Also, while the TV series is not generally available on DVD or via streaming services, intrepid web-searchers can often find entire episodes posted on the Internet.

---

conventional sense—the central figure in the story, and interpersonal conflicts are as likely to drive the plots as military ones." p. 79

[4] Internet Movie Database (IMDb), www.imdb.com/title/tt0041996/trivia/

[5] For example, David Barno and Nora Bensahel write: "Winning a big war from the air may require the same kind of bloody-minded Air Force leadership and stoic resolve not seen since the peak of deadly air combat in World War II. The staggering losses of the strategic bomber community flying against German defenses over Europe in 1943 were strikingly portrayed in the classic movie Twelve O'Clock High. [...] Screening the film and discussing its meaning throughout the force might be one small step towards bolstering the Air Force profession of arms for a war where many airmen may not return to fly and fight again." See "The Catastrophic Success of the U.S. Air Force." *War on the Rocks.* May 3, 2016. warontherocks.com/2016/05/the-catastrophic-success-of-the-u-s-air-force/

[6] Erickson, Glenn. "Twelve O'Clock High–Fox Reissue" *DVD Savant*, 2007, www.dvdtalk.com/dvdsavant/s2324high.html

As my obsession with the franchise grew, I searched for a personal copy of the 1948 novel on which "Twelve O'Clock High" was based. Unfortunately, the title seems to be out of print. Algorithm-driven Internet booksellers often quote prices of $100 or more, even for paperback reprint editions produced in the 1980s. Still, ownership is not impossible. Via an on-line small business specializing not in books but in "vintage" items, I found and purchased a 25-cent 1949 Bantam Books mass-market edition ("complete and unabridged"!) for only $10.

(Why all this practical info regarding how to watch or read *Twelve O'Clock High*? Because, in order for lessons to be replicable, notes on resources and methods must be shared, too.)

On the glossy front cover of my paperback edition, the pulpy illustration depicts Gregory Peck in the cockpit of a B-17 bomber. He wears a leather flying helmet, and googles pushed up on his forehead. His hands are on the plane's control column. He gazes at the reader. Over his shoulder, flaming and smoking aircraft are visible through the bullet-perforated windscreen.

Along the top edge of the cover reads this teaser question: *"Was this soldier <u>too</u> daring?*

Interlude

*In a shop window
a grin of recognition:
"Target for today?"*

Act II

The movie "Twelve O'Clock High" opens in 1949, five years after the end of World War II. An American lawyer, Harvey Stovall, is shopping for souvenirs, apparently about to return home from vacation in England. Wearing a suit and tie, walking along the street, his eye catches sight of a beaten-up "Toby Jug" in a curiosity shop's window. The vessel is shaped as the bust of a classic masked character, Robin Hood. For some reason, Stovall seems to prize the object. He excitedly purchases the chipped ceramic, brushing off the storekeeper's suggestions of more-precious items. He asks the proprietor to package the piece very carefully.

Stovall next travels by train first to Archbury Station, then by bicycle to a nearby meadow. He props his bicycle on a fence, and walks into the tall grass. A small herd of cows grazes nearby. Further on, the grass turns into a partially paved area, now in disrepair. Stovall surveys the landscape, becoming lost in memory.

The camera moves along to consider first an old windsock, and then the word "Archbury," which has been spelled out in stones arranged on the ground. The camera tracks past an air-control tower, and some decrepit buildings and sheds beyond. *"Bless 'em all ... bless them all ..."* Choruses of old tavern songs are heard on the wind. *"We are poor, lost sheep, who have lost our way ..."*

Stovall walks further out on the pavement. He takes off his glasses, and fogs them with his breath. As he buffs the lenses with his handkerchief, he looks up to the sky. The droning sounds of multi-engine aircraft are heard. The camera slides off Stovall, sideways into the windblown grass, and then upward to the heavens.

A quick dissolve puts B-17 bombers in our sky. They are preparing to land. The breezes turn into the blast of prop-wash, and the camera returns earthward to reveal an operational airfield. It is now 1942—early in the Americans' involvement in the war. Headquartered at Archbury field, the U.S. Army Air Forces' 918th Bomb Group is taking heavy losses, morale is low, and performance metrics are flagging.

Interlude

*He's a first-rate guy
who's gonna bust wide-open
thinking of his boys.*

Act III

Flight surgeon Capt. Donald "Doc" Kaiser notes that 28 men—three times more than normal—have requested to be medically excused from the 918th Bomb Group's next mission. "They gave reasons—colds mostly. And most haven't got colds. They haven't gone yellow. They're getting their bellies full. Do I OK them and mark them 'duty'? How much can a man take? What's physical?" he asks.

Just as frustrated as his chief physician, the group's commander Col. Keith Davenport replies: "The rules say a man goes unless he'll endanger his crew." He admits, however: "I don't know what that means."

Because his emotional connection with his men is thought to be clouding his judgment, Davenport is soon relieved of command of the 918th. The change comes at the recommendation of his friend, Brig. Gen. Frank Savage. It is not an outright demotion; Davenport continues to be well-regarded by his superiors. He is moved up to air-headquarters staff, while Savage is moved down. While Davenport initially resents the move, the two remain friends and colleagues. Savage is now himself charged with turning around the unit's culture and performance, and with helping to prove that the American practice of daylight precision bombing can be effective.

To do so, he will have to push himself, his men, and his organization to a near-breaking point.

"I guess I don't have to tell you what's coming, Frank. I'm promising you nothing except a job no man should have to do who's already had more than his share of combat," Maj. Gen. Patrick Pritchard tells Savage. "I've gotta ask you to take nice kids and fly them until they can't take any more, and then put 'em back in and fly 'em some more. We've got to try to find out just what a maximum effort is. How much a man can take and get it all."

He sighs, and continues: "I don't even know if any man can do it. That's what cracked Keith."

Interlude

*Those things are coming—
Replacements ... Combat limits ...
Right now, we hang on.*

Act IV

I first learned the language of "lessons-learned" in week-long classroom course at the Center for Army Lessons Learned ("CALL") at Fort Leavenworth, Kansas. There, we learned that a "lesson" is defined as "knowledge based on experience," and that a "lesson-learned" is "knowledge based on experience that is used to change organizational or individual behavior."[1]

A third part is key to creating organizations that can adapt to change: Lessons must be shared. In CALL jargon, a lesson-learned is "integrated" when it is shared—so that others can learn from our mistakes, and our successes.

(I have found these concepts applicable to nearly everything I have written since, including poetry. *Everything* has potential to be a lesson.)

In the late 2000s, my temporary job title was "Lessons-Learned Integration (L2I) Analyst." Armed with our civilian journalism skills and the tools we learned at CALL, our team wanted to identify whenever Iowa soldiers overcame challenges while preparing for deployment to Iraq and Afghanistan. And we wanted to share those "how-to" techniques and insights with others.

We always started with a given unit's assigned mission statement: The "Who," "What," "Where," and "Why" of any organization. From there, we moved on to consider the "Whens" and "Hows" of specific tasks and training events: *What did you set out to do? How did things turn out? What would you do differently, if you did it all again?*

As a handy framework for presenting our observations, we used the mnemonic acronym "DOTMLPF" (pronounced "dot-mil-pee-eff").

---

[1] U.S. Department of the Army. (2017). *Army Lessons Learned Program.* Army Regulation 11-33 (AR 11-33): The Army defines "lesson" as "a potential solution to a problem experienced as a result of an observation," while a "lesson learned" is a "an implemented corrective action which leads to improved performance or an observed change in behavior." p. 22

According to Uncle Sam,[2] any problem-and-solution pairing can be described in terms of one or more factors:

- **Doctrine:** How the organization "thinks" and acts within the published principles of its larger institution. In military terms: *How do we fight?*
- **Organization:** How is the organization configured? *Who answers to whom?*
- **Training:** How does the organization instruct its members and maintain their proficiencies toward performing individual and collective tasks?
- **Materiel:** How is the organization equipped? Weapons, vehicles, clothing, rations, etc.
- **Leadership & Education:** How are individual personnel managed, and how are they institutionally schooled?
- **Personnel:** What people are doing which jobs, both officially and unofficially?
- **Facilities:** What positive and/or negative effects are caused by physical infrastructures, including buildings, training areas, roads, etc.?

Viewed through the lens of DOTMLPF, the movie "Twelve O'Clock High" offers a crash-course in how to observe and apply lessons-learned. That's not by coincidence.

When U.S. Army Air Forces (USAAF) Brig. Gen. Frank Savage first arrives to his new command, for example, the gate sentry waves at the staff car without checking identifications. Savage orders his driver to stop, and exits the vehicle in order to chew out the guard personally. The memorable scene is based on screenwriter Sy Bartlett's own wartime experiences. While serving as a staff officer in VIII Bomber Command, he one day accompanied Brig. Gen. Frank Armstrong on a surprise visit to an airbase called Thurleigh. "When we drove through the gate,"

---

[2] U.S. Department of Defense. (2018) Chairman of the Joint Chiefs Instruction (CJCSI) 3010.02E dated Aug. 17, 2016 and marked current as of Aug. 16, 2018. This document defines each DOTMLPF factor, and how each relates to implementations across multiple U.S. military branches.

Bartlett recalled, "past the sentry without being challenged, I knew it was going to be one of those days. Frank [Armstrong] blew his stack. It was only the beginning. Thurleigh was a mess. The officers were drunk. All over the base military protocol and discipline were completely lax. There was no pride whatsoever. Frank found it appalling."[3]

As eventually depicted in Bartlett's screenplay, which was co-written with pilot and fellow U.S. officer Beirne Lay Jr., the incoming commander of the fictional 918th Bomb Group applies his own observations to make concrete changes at the airbase called Archbury. Savage makes changes in duty assignments, daily training, and even sleeping arrangements. He demotes the Air Operations officer, whom he suspects of cowardice in addition to indiscipline, and places him personally in charge of a problematic air crew. (Observed DOTMLPF category: "Personnel.") He takes the squadron temporary off-line, and begins drilling air crews daily in flying and bombing fundamentals. ("Training.") When he observes pilots making life-and-death decisions in the air based on their friendships with roommates on the ground, he even orders all billeting assignments be changed. ("Facilities.")

Regardless of age, era, or wartime service, each of these techniques should sound familiar to military veterans. "It had an authenticity seldom seen in war movies," *Air Force Magazine's* John T. Correll writes about "Twelve O'Clock High."[4] "It pushed all the right buttons for airmen, who held it in such regard that the movie became something of a cult film for several generations of Air Force members." Correll was a former editor-in-chief of the magazine, and served in the Air Force himself between 1962-1982. "In those days, almost everybody in the Air Force had seen it at least once, and the film was used for many years in USAF leadership courses."

Here's another endorsement regarding the movie's accuracy: "Everything [Sy] Bartlett and [Beirne] Lay ever wrote were facts," said

---

[3] Rubin, Steven J. *Combat Films: American Realism, 1945-1970.* Jefferson McFarland, 1981. p. 129
[4] Correll, John T. "The Real Twelve O'Clock High" *Air Force Magazine* Jan. 1, 2011. www.airforcemag.com/article/0111high/

John deRussy, operations officer for the real-life 305th Bomb Group during the war, and later a technical advisor on the film. "But not all of it happened in the same organization."[5]

For some, the leadership lessons delivered by "Twelve O'Clock High" may seem old-school and antiquated. For others, the tensions depicted between the methods and styles of two principal characters—the empathetic Col. Keith Davenport and the hard-charging (and possibly toxic) Brig. Gen. Frank Savage—reveal truths that are potentially as applicable today as they were during World War II.

In a "Top Ten" list generated by an informal survey of *Inc.* magazine readers in 2000, Leigh Buchanan and Mike Hofman write: "'Twelve O'Clock High' will not appeal to hierarchy flatteners and empowerment enthusiasts trying to build cultures as warm and welcoming as the family den. Savage succeeds because he doesn't give a damn about his own popularity—only about the effectiveness of the squadron. And no, good leaders don't always have to be sons of bitches. But Savage would argue that when the straits are dire and the stakes enormous, it's the only way."[6]

In a 2001 case-study curriculum developed by Hartwick College's Humanities in Management Institute, the authors offer seven different theoretical frameworks through which to discuss the management behaviors depicted in "Twelve O'Clock High." Each provides a potential language or logic—sort of a business-school version of "DOTMLPF" methodology—for analyzing the movie's events. For me, their most-useful lens regards "transformative power": the ability of an individual to translate an idea into reality and to sustain it.[7]

---

[5] Duffin, Allan T., and Paul Matheis. *The 12 O'Clock High Logbook: The Unofficial History of the Novel, Motion Picture, and TV Series*. Kindle ed. BearManor Media, 2005

[6] Buchanan, Leigh and Mike Hofman. "Everything I Know about Leadership, I Learned from the Movies" *Inc.* magazine on-line. March 1, 2000. www.inc.com/magazine/20000301/17290.html

[7] U.S. Department of the Army. (2019) *Army Leadership and the Profession*, Army Doctrine Publication 6-22 (ADP 6-22), section 1-73 defines "Leadership" as "the activity of influencing people by providing purpose, direction, and motivation to accomplish the mission and improve the organization." Of course, modern wags might also relate the concept of transformational leadership to a "reality distortion filter"—a

Years later, I realize that this is the central engine of lessons-learned—indeed, much of journalism: We like to tell stories of humans who are struggling to transform ideas into reality. Whether you're building a house or building a team or even fighting a war, the same basic questions apply: *What did you set out to do? How did things turn out? What would you do differently, if you did it all again?*

For the Hartwick curriculum developers, despite the character's apparent heartlessness, Frank Savage is an archetype of leadership likely to appeal to modern audiences. After all, Savage is a leader who says what needs doing, and then does it. "Savage clearly is not into keeping with the present times in terms of leadership style, but if the film does the trick, then every participant, whether they like Frank Savage or not, will admire him as a leader. [...]" they predict.[8]

"In the space of a short film, people can dislike Savage, but choose to follow him. Why? Because they believe in the mission as he articulates it. Isn't that transformational? If Frank Savage can gain a following [today], then truly he is a timeless, transformational leader."

What did you set out to do? How did things turn out? What would you do differently, if you did it all again?

---

term first employed to describe the leadership presence and persona of Apple Computer co-founder Steve Jobs.
[8] Insigna, Richard. *Teaching Notes for Twelve O'Clock High*. Hartwick Humanities in Management Institute, 2001. p. 14

Interlude

*Take nice kids and fly
until they can't anymore.
And then, fly some more.*

Act V

Sy Bartlett and Beirne Lay Jr. each enjoyed a privileged position from which to observe the air war over Europe. They were motivated to document history, and were each uniquely qualified and located to do so. And yet, working together after the war, they struggled to package their World War II experiences into an effective narrative payload. Reportedly, their first attempts were all over the map.

Lay had enlisted in the U.S. Army 1932-1935. As a civilian during the "interwar" years, he wrote journalism, memoir, and opinion articles about air power. He returned to active-duty in 1939. (His first screenplay, "I Wanted Wings," was produced as a 1941 movie.) Back in uniform, he was first a flying instructor, and later a speech writer and historian. Lay served as one of six original staff officers under Col. Ira C. Eaker.[1] In 1942, the team was dispatched to England in preparations for creating a U.S. strategic bombing effort from the ground up.

Eaker and his small team were tasked with learning everything they could from the British about strategic bombing. That included a certain amount of hands-on learning: "We won't do much talking until we've done more fighting," Eaker graciously told his British hosts.[2] Eaker was soon promoted to brigadier general, and assigned to himself command American air forces in Europe. First called the U.S. VIII Bomber Command, the organization was later labeled the U.S. Eighth Air Force.

During this period, Lay co-piloted a handful of missions on B-17 "Flying Fortress" bombers. Based on those experiences, he wrote a staff analysis about the Schweinfurt-Regensburg raids. The guts of that report would become a 1943 article in the *Saturday Evening Post*,[3] and also a

---

[1] Miller, Donald L. *Masters of the Air: America's Bomber Boys Who Fought the Air War Against Nazi Germany* Kindle ed. Simon and Schuster; New York, New York. 2006
[2] Freeman, Roger A. *The Mighty Eighth: Units, Men and Machines*. Doubleday and Company, Inc. Garden City, New York. 1970. p. 6
[3] Lay Jr., Lt. Col Beirne "I Saw Regensburg Destroyed" *Saturday Evening Post*, Nov. 6, 1943. Lay was a prolific writer, both in and out of uniform. Regarding the Schweinfurt-

chapter in the 1948 novel *Twelve O'Clock High*. Later, he took command of a stateside unit of B-24 "Liberator" bombers, with which he trained and returned to Europe. He was shot down over France in 1944, but evaded capture until the allies invaded at Normandy.

Lay's post-war writing partner Sy Bartlett was also a strong advocate of air power. Bartlett, a former journalist and Hollywood screenwriter, had joined the military as a U.S. Army Signal Corps officer. First assigned to make Army training films, he later talked himself onto the U.S. VIII Bomber Command's operational intelligence staff. He was put in charge of a signals intelligence team, and flew as an observer on both American and British bombing raids.

Bartlett was not above self-promotion. As an observer on the first American bombing mission over Germany—a raid on Wilhelmshaven by the U.S. 306th Bomb Group—Bartlett reportedly jostled the bombardier's release button by accident, dropping the B-17's payload early and off-target. As such, he subsequently claimed have been the first American to drop bombs on Germany. He later purposefully repeated the action while observing an RAF bombing raid on the German capital in March 1943, and called a press conference boasting that he was the first American to drop bombs on Berlin.[4]

In *Combat Films: American Realism 1945-1970*, Stephen Jay Ruben quotes interviews with Bartlett and Lay about "Twelve O'Clock High." According to Ruben, Bartlett and Lay first discussed co-writing an "intimate narrative about the Eighth Air Force" when the two friends met briefly at the Thurleigh airbase.[5] The meeting took place following Lay's extraction from Occupied France. The location is also notable, given that Thurleigh would become the inspiration for Archbury, the fictional base used in "Twelve O'Clock High." Lay was about to ship

---

Regensburg mission, see also Lay's "What It Takes to Bomb Germany: The Kind of Men Who Can Do It" *Harper's* magazine, November 1943

[4] Duffin, Allan T., and Paul Matheis. *The 12 O'Clock High Logbook: The Unofficial History of the Novel, Motion Picture, and TV Series*. Kindle ed. BearManor Media, 2005

[5] Rubin, Steven J. *Combat Films: American Realism, 1945-1970*. McFarland, 1981. pp. 126-128

back to the United States—as a successful escapee, he now knew too much about the French Resistance to again risk capture flying over Europe. He would get another command and later fight in the Pacific Theater, however, flying B-29 "Superfortress" bombers. At that final wartime meeting at Thurleigh, Bartlett reportedly told him:

"You know, when this war ends, people are going to forget what happened here. They won't care anymore. To prevent that from happening, you and I are going to write a novel about the Air Corps. [...] And when we finish the novel, we'll write a screenplay. It will be the best war film ever. With your combat experience and my inside knowledge of the command structure, we can write this thing from a unique point of view."[6]

What did you set out to do? How did things turn out? What would you do differently, if you did it all again?

---

[6] Rubin, pp. 126-128

Interlude

*There's only one hope:
daylight precision bombing
will shorten the war.*

Act VI

As a story, write Allan T. Duffin and Paul Mattheis, the book *Twelve O'Clock High* presented "a mix of combat heroics, interpersonal dramas, and romantic affairs. It was a potboiler with a military edge."[1] However, while aimed at commercial success—writing for economic gain, after all, is the very definition of "potboiler"—*Twelve O'Clock High* was a potboiler with a purpose.

And that purpose, arriving as it did against a Cold War backdrop of atomic bombs, anti-Communist witch hunts, and pending American involvement in the defense of Korea, was to tell a foundational story of the U.S. Air Force. It helped establish a mythic narrative regarding strategic bombing, which still echoes in air-power discussions today.

Partially based on true events and personalities in the U.S. Army Air Force's 306th Bomb Group, the movie "Twelve O'Clock High" focuses on the fictional 918th Bomb Group, during the early years of American involvement in World War II. In the true-to-life narrative, American air generals are desperate to prove their young service's value by contributing to the war effort at strategic level, alongside the nation's ground and sea services.

In 1941, the U.S. Air War Plans Division plan No. 1 (AWPD-1) was generated by a small task force of forward-thinking U.S. Army Air Corps planners, each of whom had previously instructed at the Air Corps Tactical School (ACTS) at Maxwell Field, Alabama. While there had been plenty of journal articles written in the period between the world wars, the AWPD-1 was a codification of the U.S. military's thinking about how air power should be used. Part of an informal intellectual movement of Army aviators called the "Bomber Mafia," the editorial group comprised Col. Harold L. George, Maj. Laurence S. Kuter, Lt. Col. Kenneth Walker, and Maj. Haywood S. Hansell Jr. Each would go on to make other great contributions to U.S. Air Force history. (All but

---

[1] Duffin, Allan T., and Paul Matheis. *The 12 O'Clock High Logbook: The Unofficial History of the Novel, Motion Picture, and TV Series*. Kindle ed. BearManor Media, 2005.

Walker survived World War II; Walker was awarded the Medal of Honor for actions in the Pacific.) The plan briefly and broadly delineated efforts in U.S. military aviation that would be required in a war against Germany, Italy, and Japan. Hansell would later write:

"Harold George made the decision to adopt this approach: A strategic air offensive to debilitate the German war machine and topple the German state if possible, and to prepare for support of invasion. [The AWPD-1 plan] did not eliminate the possibility of achieving victory through strategic air warfare alone, or limit the effort of that objective. It placed prime emphasis on air power, including strategic bombardment. But it also acknowledged that an invasion might be necessary and an air offensive was essential in preparation for such an invasion and subsequent combined operations."[2]

In "Twelve O'Clock High," the narrative focus is on efforts to prove the validity of "high-altitude daylight precision bombing." The technique requires air crews to visually acquire a target through clouds, smoke, and fog, while buffeted by anti-aircraft fire and fighter-plane attacks. On approach, bombardiers set their super-secret Norden bombsights—analog computers that calculate factors such as airplane direction and speed, wind direction and speed, atmospheric conditions, and type of ordnance. With controls connected to the bombsight, the planes steer on autopilot for the final seconds prior to bomb-release.

Reflecting such techniques and technologies, American mission-planners target industrial "choke points" in the German economy, in order to cut-off war production and transportation. Examples of such targets included ball-bearing factories, synthetic-oil refineries, and railroad network hubs. Such attacks, planners argue, will minimize civilian casualties and will achieve the following strategic goals:

- Decrease enemy capacities and resources to replace fighters, bombers, and other war equipment.
- Work toward a condition of air superiority—dominating the skies—that will be necessary before launching any successful allied ground invasion of the European continent.

---

[2] Haywood S. Hansell Jr., *The Air Plan that Defeated Hitler*, Arno, 1980. pp. 75-76

Early in the war, key assumptions of precision-bombing doctrine included:

- Heavy bombers such as the B-17 "Flying Fortress" and B-24 "Liberator" can fly higher and faster than interceptor aircraft and anti-aircraft fire. In the words of British prime minister Stanley Baldwin: "The bomber will always get through."[3]
- Bombers can defend themselves without fighter escorts. (At the beginning of American involvement in the air war, fighters' effective ranges maxed out at the German borders.)
- Technologies such as the Norden bombsight can deliver ordnance, under combat conditions, with sufficient precision to avoid collateral civilian damage. In one particularly resilient phrase, proponents boasted that U.S. bombardiers "could hit targets the size of a pickle barrel."[4]

Unfortunately, many of these assumptions proved variously unwise or untrue. Once U.S. bombers flew beyond fighter-escort ranges, for example, German fighter planes and anti-aircraft fire ("flak") caused heavy U.S. losses. And, given the limitations of the best technologies of the day, "precision-bombing" still wasn't very precise. In reality, "precision" never got better than a circle of average probability approximately three-quarters of a mile in radius.[5] In short, the daylight precision bombing concept was arguably more myth than reality. It wasn't until Vietnam War, or perhaps even Operation Desert Storm, that the promises of precision bombing were available and realized.[6]

---

[3] Baldwin, Stanley. "A Fear for the Future." Speech delivered to British House of Commons Nov. 10, 1932. Center for Strategic and International Studies. missilethreat.csis.org/wp-content/uploads/2020/09/A-Fear-for-the-Future.pdf

[4] Correll, John. T. "Daylight Precision Bombing" *Air Force Magazine*. www.airforcemag.com/article/1008daylight/

[5] Miller, Donald L. *Masters of the Air: How the Bomber Boys Broke Down the Nazi War Machine*. Kindle ed. 2020. "Average bombing accuracy over the course of the war, 'expressed as a circular error probable,' was approximately three-quarters of a mile, hardly the circumference of a pickle barrel."

[6] Correll, "Daylight Precision Bombing": "Precision guided munitions first gained fame in the Vietnam War, but it was in the Gulf War and other conflicts of the 1990s that

The Americans' commitment to daylight precision bombing in World War II—at least philosophically, if not always in practice—differed from the British techniques of nighttime "area bombing." Rather than choke off industrial production, area-bombing missions sought to demoralize and terrorize the enemy's civilians. Area bombing was less precise, but was also less susceptible to weather hazards—one didn't need to see into a fog to bomb through it.

The British had started the war with a precision-bombing approach of their own, but lost too many men and machines. They urged the Americans to join them in conducting night raids. The American air generals, however, feared being subsumed under British command.

Only a short memo and a rhetorical flourish by U.S. Eighth Army Air Force Maj. Gen. Ira Eaker saved the day. At a January 1943 conference of allied leadership at Casablanca, Morocco, Eaker convinced British Prime Minister Winston Churchill to hold off on a request to have the Americans join the British in bombing at night. Armed with a one-page memo—Churchill valued brevity—Eaker pointed out the potential for around-the-clock allied bombing of Germany. "The devils will have no rest," he assured the prime minister.[7]

Often, nighttime raids featured the use of incendiary bombs to burn people out of their homes. Later, the Americans would also participate in nighttime raids against Germany.[8] This notably included the February 1945 firebombing of Dresden, which killed approximately 25,000 and

---

the Air Force finally achieved pickle barrel accuracy, placing bombs within 10 feet of the aim point. The use of the Global Positioning System and satellite data for aiming had made the issue of day vs. night irrelevant."
[7] Wolk, Herman "Decision at Casablanca" *Air Force Magazine*, Jan. 1, 2003. www.airforcemag.com/article/0103casa/
[8] Correll, "Daylight Precision Bombing": "From 1942 on, 56 percent of [RAF Bomber Command] sorties were against cities. On some occasions, notably the bombing of Dresden in 1945, the [U.S. Army Air Force] joined the British in bombing cities, but overall, less than four percent of U.S. bombs in Europe were aimed at civilians. The main targets for the AAF were marshaling yards (27.4 percent of the bomb tonnage dropped), airfields (11.6 percent), oil installations (9.5 percent), and military installations (8.8 percent)."

displaced between 100,000 to 200,000 people.[9] (An American soldier taken prisoner-of-war named Kurt Vonnegut survived that bombing, and would later write a satirical novel partially set in that horrific event. The book *Slaughterhouse-5* was named after the brick structure that housed Vonnegut and other U.S. prisoners overnight, inadvertently protecting them from the flames.)

In 1942-43, however—the contexts in which the narratives of "Twelve O'Clock High" and its first seasons as a TV show take place—the Americans were conducting a maximum effort with minimal personnel and equipment, while seeking to prove the validity of daylight precision bombing. It was a brutal time, in its own way. In the abstract, U.S. bomber crews were able to request reassignment after completing 25 missions. In 1943, however, the average life-expectancy of a crew member was 11 missions. Chances of surviving 25 missions were 1 in 4.[10] Additionally, in a 1944 study of crew members who successfully achieved 25-missions, all reported one or more symptoms of "combat fatigue"—what we now would label post-traumatic stress.[11]

In his 1961 satirical novel about B-25 "Mitchell" bomber crews conducting missions over Italy during World War II, former U.S. Army Air Corps bombardier Joseph Heller would famously label such an apparent no-win calculus as a "Catch-22." A crew member cannot request to be taken off flight-duty because he believes himself to be insane, because such a request indicates that he is sane. Where Heller's protagonist bombardier Capt. John Yossarian faces the absurdity of impossible odds with heavy irony, however, "Twelve O'Clock High" presents the question of survival as a matter-of-fact challenge to be overcome, both individually and collectively. "Forget about going home," Brig. Gen. Frank Savage goads the men of 918th Bomb Group,

---

[9] Miller: "Two-thirds of the men could expect to die in combat or be captured by the enemy. And 17 percent would either be wounded seriously, suffer a disabling mental breakdown, or die in a violent air accident over English soil."
[10] Miller
[11] Miller

echoing the stoicism of Marcus Arelius.[12] "Consider yourselves already dead. Once you accept that idea, it won't be so tough."

Savage, of course, is fighting a war larger than himself. His biggest enemy is time. He needs time for his leadership methods to revive unit morale and pride. Time enough to prove the value of daylight precision bombing. And time enough to see the promised flood of reinforcements—both planes and trained air crews—finally arrive in theater.

"[W]e're guinea pigs, Jesse. But there's a reason," the general tells Lt. Jesse Bishop, a heroic young pilot who is struggling to find purpose in either his mission or the sacrifices of his comrades. "If we can hang on here now, one day soon, somebody's gonna look up and see a solid overcast of American bombers on their way to Germany to hit the Third Reich where it lives. Maybe we won't see it. I can't promise that. But I can promise they'll be there, if only we can make the grade now."

These considerations notably ignore some big-picture moral questions, including whether or not the targeting of civilians is ethically defensible. Or whether it is morally correct (or "sweet and fitting," to borrow from World War I poet Wilfred Owen)[13] to die for one's country. Instead, Savage simply urges Bishop to fight on—he promises there is a new day coming, even though they each may not live to see it.

To evoke Victorian poet Alfred, Lord Tennyson:[14]

[...] Theirs not to make reply,
Theirs not to reason why,
Theirs but to do and die.

---

[12] Arelius, Marcus. *Meditations.* Book 7, verse 56. "Think of yourself as dead. You have lived your life. Now take what's left and live it properly."

[13] Owen, Wilfred. 1921. "Dulce et Decorum est," first published in a posthumous collection, *Poems.* Chatto & Windus, 1920. p. 15. Also: www.poetryfoundation.org/poems/46560/dulce-et-decorum-est

[14] Tennyson, Alfred Lord. "The Charge of the Light Brigade," first published in *The Examiner* newspaper, Dec. 9, 1854. Also: www.poetryfoundation.org/poems/45319/the-charge-of-the-light-brigade

Interlude

Forget about home.
Once you accept that you're dead,
it won't be so tough.

Act VII

In the 2016 superhero movie "Deadpool," the buffoonish title character repeatedly exposes himself to horrific injury only to comically bounce back—always, with a sarcastic quip. In one notable scene, as he steps off a highway bridge to stop an armored convoy that is fast-approaching below, he seems to steel himself by muttering an enigmatic catchphrase: "maximum effort."[1]

From context, the phrase seems to refer to pushing oneself past physical limitations and psychological pain. It is an echo from a million high-school half-time pep-talks, a mix of "mind over matter" and "leave it all out there on the field." Fans of old war movies, of course, will likely recognize the phrase.

The phrase "maximum effort" did not originate with *Twelve O'Clock High*. For example, in a 1941 RAF-produced documentary film "Target for Tonight," the phrase is used to describe an all-out commitment of available bomber assets for the night's mission. "Bomber Command to all groups," the field order goes out via scrambled voice-telephone message. "Target for tonight ... Maximum effort tonight ... Town 4-3-4 [Kiel], Germany ... Target: Naval docks and barracks."

The phrase similarly shows up in other U.S. documentary and feature film footage from the 1940s and '50s, as well as U.S. Air Force white papers in the 1960s and '70s.

At the unit level, "maximum effort" means deploying 100-percent of available strength (planes and crews) without regard to cost or resources. (Contrast this with the principle of war called "economy of force," which current U.S. doctrine defines as "The judicious employment and

---

[1] The ultrafast-healing mutant mercenary Deadpool is an individual, but the phrase "Maximum Effort" can also be applied to organizations. Ryan Reynolds, the Canadian actor-producer who spent years lobbying to bring the red-and-black-suited anti-hero to the silver screen, now owns a real-life marketing and production company called "Maximum Effort." Its mission statement? "We make movies, TV series, content and cocktails for the personal amusement of Hollywood Star Ryan Reynolds." See: www.maximumeffort.com/aboot

distribution of forces so as to expend the minimum essential combat power on secondary efforts to allocate the maximum possible combat power on primary efforts."[2])

At the operations level, "maximum effort" means launching a maximum number of squadrons and groups, with the intent of overwhelming enemy air defenses.

At the strategic level, "maximum effort" means demonstrating, to the enemy and allies and even the taxpayers back home, that daylight precision bombing is an effective tool in degrading enemy industrial production.

Where "Twelve O'Clock High" achieves its catalytic genius, however, is when the character of Maj. Gen. Patrick Pritchard—the commander above Brig. Gen. Savage—applies the phrase "maximum effort" at a *human* level: "We've got to find out just how much a man can take."

Within the narrative of "Twelve O'Clock High," then, the question of how to produce "maximum effort" is as much a psychological test as it is a physical one. Leaders must motivate airmen to risk their lives, mission after mission, day after traumatic day. Individual airmen must fight through flak and fighters during hours-long missions, breathing through oxygen masks while operating in the high-altitude, frost-bite cold. After "bombs away," they whip-saw back to the relative safety and security of their air bases in England, in order to patch-up themselves and their planes, and to prepare to fly their next missions.

It is no wonder that the U.S. military's institutional appreciation of Post-Traumatic Stress began to evolve in World War II.

Just how much war, after all, can someone take?

---

[2] Department of Defense, "Joint Operations" Joint Publication 3-0 (JP 3-0), Appendix A, p. 2. Also:
www.jcs.mil/Portals/36/Documents/Doctrine/docnet/jp30/story_content/external_files/jp3_0_20170117%20.pdf

Interlude

*Every deadbeat
gets assigned to a plane called
"Leper Colony."*

Act VIII

Brig. Gen. Frank Savage moves to command the 918th Bomb Group with a footlocker full of tough-love. He strictly enforces standards, procedures, and protocol. To many, his authoritarian methods seem abusive. Maybe they are.

For example, any crew member who fails to make the grade is loudly and publicly assigned to a single crew of ne'er-do-wells, upon which the commander's scorn seems particularly focused.

Accusing him of cowardice, Savage demotes Maj. Ben Gately to serve as the pilot in command of a plane labelled as "The Leper Colony." Gately, the privileged West Point graduate son of a general, protests at first, but then withdraws his remarks when Savage threatens to let Gately complain directly to headquarters.

Later, Gately defensively introduces himself to his new crew: "In case you aren't clear what this is about, I'm supposed to be a deadbeat. So are you. That's why you were assigned to me. 'The Leper Colony.' How do you like it? You'll like it less after a mistake. A blowtorch is turned our way. Nobody's shoving me into it. Is that clear?"

Gately's experiences will prove to be a crucible.

Maj. Harvey Stovall is a veteran of the first World War and a lawyer in civilian life. After first coolly receiving the unit's new commander, Stovall gradually comes to realize that Savage is just as emotionally committed to his men as was the former commander, Davenport. Savage is just better at hiding it. As the unit's adjutant, Stovall is responsible for administration of military paperwork, including requests for personnel transfer. Behind the scenes, he becomes a co-conspirator with Savage, bending the rules and delaying reassignments in the hope that unit cohesion will improve after Savage delivers a few concrete victories.

Interlude

Stay in formation.
What is not expendable
is your loyalty.

Act IX

As they evolved, daylight precision bombing tactics called for various squadron and group formations, which changed in size, number and configuration. Regardless of size or timeframe, however, formations were intended to optimize interlocking fields of defensive machine-gun coverage, group maneuverability, and accuracy in bombing.

In the calculus presented in "Twelve O'Clock High," the failure to successfully launch just one B-17 "Flying Fortress" represents the subtraction of 17 percent of a 6-plane squadron's combat power over target. The loss of one B-17 means the loss of 10 trained personnel, along with the defensive firepower they each provide: pilot, co-pilot, navigator, bombardier, radio, top-turret gunner, ball-turret gunner, tail-gunner, and two waist-gunners. Reports to headquarters show that the 918th Bomb Group is regularly failing to put up the maximum number of crews and planes.

"Every gun is able to give the group maximum defensive firepower," Brig. Gen. Frank Savage tells his officers. "When you go out of formation, you reduce the group's power by 10 guns. A crippled airplane is expendable. What is never expendable is your obligation to this group. This group. This *group* ..."

Gradually, Savage's enforcement of standards seems to turn the unit around. Combat losses decrease, presumably due to improved discipline while flying in formation. Morale further improves when Savage himself risks charges of insubordination on behalf of his unit, claiming that a radio malfunction prevented him from hearing a weather-related order to scrub a mission. While other units dutifully returned to base, he defiantly observes to Maj. Gen. Patrick Pritchard, the 918th went on to hit their targets. He brazenly even requests a commendation for the action—not for himself, but for his unit.

The 918th Bomb Group soon hits an operational rhythm. Heading into the fictionalized 1944, the practice of daylight precision bombing seems to be validated. Targets move deeper and deeper into enemy-held territory. When it comes time to target Germany itself, unit morale is so high that even non-combatant ground personnel—including the chaplain and unit adjutant—stowaway aboard various planes to participate in the attack as waist-gunners. Upon the mission's return, Savage catches the staff officers sneaking away from the aircraft. After Savage chews them out for being reckless, adjutant Maj. Harvey Stovall quips that, while his glasses were fogged up, he still thinks "he got a piece of one of them."

"Ours or *theirs?"* Savage snorts, angrily.

Interlude

What happened to him?
State of shock. Complete collapse.
He's up there, with them.

Act X

Over time, the pace of operations begins to wear down even the stoic Brig. Gen. Frank Savage. Maj. Gen. Patrick Pritchard tells his staffer Col. Keith Davenport—once himself the commander of the 918th—to quietly research candidates to replace Savage, in case of an emergency.

Then, it happens. After the death of a much-loved airman, Savage is suddenly unable to pull himself up into his aircraft. With the squadron's propellers already turning for take-off, the once black-sheep Maj. Ben Gately takes Savage's place as pilot of the lead plane. The unit's chaplain and doctor ease Savage back to the operations shack. After a short emotional outburst at the side of the runway, helplessly urging them to call off the mission, Savage lapses into a catatonic fugue.

While Savage sits motionless in his office, Davenport and the 918th's ground staff sweat-out the long hours. Only after the return of 19 aircraft does Savage return to lucidity. He tells his colleagues that he is very tired, and racks out in a nearby bunk. Davenport puts a blanket over Savage, and removes the latter's flying boots—gear that Savage had once given to Davenport. Loyal adjutant Lt. Col. Harvey Stovall closes the door the commander's office, and walks outside. The droning of bombers is heard overhead. He looks to the sky.

The movie returns to present-day 1949, and ends as the civilian-suited Stovall walks the tarmac back to his bike. As he pedals away from Archbury field, the soundtrack music rises. Singing voices are heard on the wind. *"We are poor lost sheep, who have lost their way ... "*[1]

Fade to black.

---

[1] Boehm, Richard Theodore "Tod B. Galloway: Buckeye Jongleur, Composer of 'The Whiffenpoof Song'" *Ohio History Journal* Vol. 83, No. 4, Autumn 1974, pp. 256-282. "The Whiffenpoof Song" is a humorous musical adaptation inspired by Rudyard Kipling's 1892 song-poem "The Gentlemen Rankers," which was published in his 1892 collection "Barrack-Room Ballads." The 1919 Whiffenpoof version is credited to American Tod. B. Galloway, who served as a uniformed YMCA entertainer of the troops in World War I.

Interlude

*We've got to find out
just how much a man can take:
"Maximum Effort."*

Act XI

When Sy Bartlett and Beirne Lay struggled to condense their potboiler novel into a workable screenplay, "maximum effort" became their touchstone. "As Lay recalled later, at one point [Producer Louis D.] Lighton asked the two authors to define the central idea of '12 O'Clock High.' We each gave him our answers," said Lay, "and he said, 'Well, Christ, fellows, you're going all over the place here. Give me something you can put down in a paragraph, or preferably in a line, or even four words." After they decided to make Savage's mental breakdown the main point of the story, Lighton agreed with them: "I don't want one inch of film that does not contribute to telling that central idea."[1]

Like an air crew tossing out the weight of unneeded equipment in order to stay in the sky, Bartlett and Lay jettisoned narrative sub-plots about wartime romances, signals intelligence, and upper-echelon politics. They focused everything they had on one central theme.

Other competitive war movies lack that focus. Take, for example, 1948's "Command Decision," a feature film released a year ahead of "Twelve O'Clock High." First a stage play and novel, and then a Metro-Goldwyn-Mayer Studios movie, "Command Decision" featured a star-studded cast of actors—Clark Gable, Walter Pidgeon, and Brian Donlevy—portraying U.S. Army Air Forces bomber-generals. Van Johnson also headlined, cast as a savvy tech sergeant who works the system to provide moments of levity and charm.

It is a fine movie, and shares much in common with "Twelve O'Clock High." Author William Wister Haines, who based his narrative on his own World War II experiences within the walls of the Eighth Air Force, offers his own unique point of view on what it takes to be a transformative leader. Some of actor Clark Gable's wartime insights may also be present. Gable was a veteran B-17 waist-gunner, and even

---

[1] Duffin, Allan T., and Paul Matheis. *The 12 O'Clock High Logbook: The Unofficial History of the Novel, Motion Picture, and TV Series*. Kindle ed. BearManor Media, 2005

produced "Combat America,"[2] a chummy 62-minute lessons-learned documentary about the experience.

But each of those works, however commendable in telling the story of the Eighth Air Force, seems to lack focus.

They lack poetry.

What they lack, in my opinion, is "maximum effort."

---

[2] "Combat America." Written by John Lee Mahin. Narrated by and featuring Clark Gable. First Motion Picture Unit, U.S. Army Air Forces. Documentary about air crew life in the U.S. Eighth Air Force. 62 minutes. Color. 1945

Interlude

They all looked alike.
All the dead had just one face;
it was very young.

Act XII

In the U.S. Army, a lesson-learned is any knowledge based on experience, that is subsequently used to change individual or organizational behavior. A lesson-learned can be "integrated" by sharing it with others—maybe as an After-Action Review (AAR), or in a newsletter. More expansively, I think lessons can be shared through popular and literary forms: novels, films, TV shows, comic books, poems—whatever gets the ideas across. Whatever gets the message through. Whatever people can remember.

People can remember poems for a long time.

I learned how to write haiku in Second Grade. Decades later, at similar ages, my kids also learned how to write poems, using the same literary rules. Haiku is a deceptively simple form: a distilled observation or experience, focused in both form and content. It is also surprisingly capable of capturing complexities, contradictions, and truths.

In translating their novel into a movie, producer Louis D. Lighton advised veterans Sy Bartlett and Beirne Lay to focus "Twelve O'Clock High" down to its essential idea: "A paragraph ... or even four words."[1]

To me, that sounds like poetry.

It strikes me that I don't now remember many of the "tactical and practical" lessons I once helped document in DOTMLPF format, as I prepared myself and other citizen-soldiers to deploy to the Global War on Terror. Our recommended techniques and procedures were constantly evolving. One day, we'd tell soldiers to drive on one side of the road to avoid Improvised Explosive Devices (IED). The next day, we'd tell them to drive on the other.

Our national purposes and objectives also shifted over time. Depending on when and where they deployed, each wave of citizen-soldiers experienced their own "maximum efforts." Remember "Shock and Awe"? "Hearts and Minds"? "Clear, Hold, and Build"? Each of these

---

[1] Duffin, Allan T., and Paul Matheis. *The 12 O'Clock High Logbook: The Unofficial History of the Novel, Motion Picture, and TV Series.* Kindle ed. BearManor Media, 2005

slogans, and the various campaigns they represent, now seems to have been written on sand. Today, I am retired from the military, the parent of children who are themselves now old enough to enlist. I find myself sifting my past for a few nuggets of wisdom to share with them—one or more lessons to share.

In thinking about Afghanistan, I often return to Archbury.

The movie "Twelve O'Clock High" is something like a morality play. It presents its protagonist's mental, emotional, physical, and organizational struggle to make his ideas and ideals a reality. The fictional Brig. Gen. Frank Savage is wholly committed to the concept of daylight precision bombing. To him, the strategic doctrine is a matter of faith. An endeavor worthy of sacrificing all: mind and body, men and equipment. *"Maximum effort."*

While Savage seems full of moral certitude, however, can we really regard his quest to be a moral one? From a movie critic's point of view, of course, it is probably unfair to extend considerations of morality beyond the narrative frame presented by the work of art itself. The ethics and effects of strategic bombing, after all, remain largely unaddressed in "Twelve O'Clock High." The movie (and the 1964-1967 TV series that would later follow it[2]) deftly avoids larger philosophical questions by keeping its focus locked on the situational imperatives of command. It seems obvious that Bartlett and Lay very deliberately set their screenplay in a very specific time and place. "Daylight precision bombing will shorten the war" is the movie's only moral fulcrum.

Still, I do find myself wondering what Frank Savage would have thought of his own actions, were we to check in on him later in his

---

[2] "The Sound of Distant Thunder" *12 O'Clock High*, created by Sy Bartlett and Beirne Lay Jr., Season 1, Episode 4, QM Productions, 1964. This is one of the few times the TV show writers wrestled with the morality of strategic bombing. The story regards a bombardier who experiences a crisis of confidence when his British love interest is killed by German bombs. See also Sam Edwards, "12 O'Clock High and the Image of American Air Power, 1946-1967," page 54: "[The '12 O'Clock High' TV program] veils its ideological commitment to the bomber through stories that ostensibly explore 'other' issues—love, trauma, Anglo-American relations. Yet the resolutions to those stories quietly assert that strategic bombing is right, proper, and effective."

fictional life. After all, later in the European air war—after the time depicted in "Twelve O'Clock High"—American air forces drifted away from the practice of precision bombing, if not entirely from the rhetoric of it.[3] While late-war strikes against oil and transportation targets may have been somewhat successful in hindering German industrial production, historians still debate about whether or not strategic bombing actually brought the war in Europe to a quicker end.

Then, there is the problem of Japan.

After taking charge of U.S. air forces in the Pacific Theater, Maj. Gen. Curtis LeMay—ironically, one of the Eighth Air Force's original squadron leaders, and upon whom the character of Frank Savage was directly based—made the decision to abandon precision bombing in the war against Japan. In the Pacific, fast-flowing winds at high altitudes known as "jet streams" made visual targeting impossible. As tailwinds, jet streams caused bombers to overshoot their targets. As headwinds, jet streams significantly slowed bombers' airspeeds and made them susceptible to anti-aircraft defenses.

Coincidentally, in taking command in the Pacific, LeMay replaced Maj. Gen. Haywood Hansell. Hansell was another of the Eighth Air Force's original squadron leaders, and also one of the "Bomber Mafia" intellectuals who wrote the Air War Plans Division No. 1 document. Based on knowledge gained from experience, LeMay ordered a change in tactics. Instead of daylight precision bombing against industrial targets, he shifted to low-altitude, incendiary missions against civilian areas.[4]

Japanese buildings were generally more flammable than European ones. In attacks that surpassed even the horrific firebombing of Dresden, Germany, Americans dropped bombs filled with jellied gasoline

---

[3] Call, Steve. *Selling Air Power: Military Aviation and Popular Culture after World War II*. College Station, Texas, Texas A&M University Press, 2009. The movie "Twelve O'Clock High" itself exists as a narrative object that restates and amplifies American air power rhetoric at the beginning of the Cold War. pp. 59; 61-62; 104

[4] The United States Bombing Surveys (PDF). Maxwell Air Force Base, Alabama: Air University Press. 1987. p. 85. Also: www.airuniversity.af.edu/Portals/10/AUPress/Books/B_0020_SPANGRUD_STRATEGIC_BOMBING_SURVEYS.pdf

("napalm") and phosphorus. Thousands of Japanese civilians died in the resulting conflagrations. One night's raid against Tokyo, a March 1945 attack called "Operation Meetinghouse," killed an estimated 100,000 civilians, and left more than one million homeless. Some 16 square miles were reduced to ash. As a single event, that night's death toll is comparable to the numbers of those killed in the atomic attacks at Hiroshima and Nagasaki in August 1945.[5]

LeMay is often quoted as saying, "If we'd lost the war, we'd all have been prosecuted as war criminals."[6]

The cover of a 70-year-old potboiler paperback laying on my desk asks, regarding Frank Savage: "Was this soldier *too* daring?" The bumper-sticker philosopher in me now asks *"WWFSD?"* regarding the inherent morality of "maximum efforts"—*"What would Frank Savage do?"* It is fun to think about in the abstract, in a conversation about movies. I shudder, however, because I suspect we already know the answer.[7]

Here's the epiphany I gained after deep-diving into all things

---

[5] Rhodes, Richard. "The General and World War III." *The New Yorker* June 19, 1995: "The United States Strategic Bombing Survey estimated that 'probably more persons lost their lives by fire at Tokyo in a 6-hour period than any time in the history of man." See also: United States Bombing Surveys, p. 92

[6] Hurley, Alfred F. and Robert C. Ehrhart (eds.) "Air Power and Warfare: The Proceedings of the 8th Military History Symposium" Office of U.S. Air Force History and U.S. Air Force Academy, Washington Oct. 18-20, 1978. In a response to a question posed to LeMay by Cadet Vance Skarstedt regarding the general's moral considerations in bombing Japan, LeMay answered, "Killing Japanese didn't bother me very much at that time. It was getting the war over that bothered me. So I wasn't worried particularly about how many people we killed in getting the job done. I suppose if I had lost the war, I would have been tried as a war criminal. Fortunately, we were on the winning side. Incidentally, everybody bemoans the fact that we dropped the atomic bomb and killed a lot of people at Hiroshima and Nagasaki. That I guess is immoral; but nobody says anything about the incendiary attacks on every industrial city in Japan, and the first attack on Tokyo killed more people than the atomic bomb did. Apparently, that was all right. [...]" p. 200

[7] Hurley and Ehrhart. In his answer to Skarstedt, LeMay concluded: "I guess the direct answer to your question is, yes, every soldier thinks something of the moral aspects of what he is doing. But all war is immoral, and if you let it bother you, you're not a good soldier." p. 201

"Twelve O'Clock High," and writing some poems in the process: There is nothing in DOTMLPF that addresses morality. Or empathy. Or how to live a good life. "A lesson-learned is knowledge based on experience, used to change behavior." Transformational leaders are neither inherently moral or immoral. Each decision—whether leader or follower—is an individual choice. A moment of truth.

The trick, I suppose, is to keep your wits and your humanity about you. Especially when you're in the middle of a fight.

You don't have to be a leader of a military unit or a business enterprise (or an armchair historian) to appreciate the dilemmas portrayed in "Twelve O'Clock High." You don't need to apply DOTMLPF or business-school frameworks of management theory. You don't have to choose between the leadership styles of Brig. Gen. Frank Savage or Col. Keith Davenport. Some of us are Maj. Ben Gately, who learns to manage his fears while focusing on leading a small crew of screw-ups that no one else seems to want. Some of us are Lt. Jesse Bishop, who just wants to figure out how he fits into the larger picture—and whether or not his sacrifices will actually make a difference.

Personally, I hope to emulate Maj. Harvey Stovall. During his time at war, he enjoys a privileged position of observation, and works behind-the-scenes to help others. The old citizen-soldier just wants to use his particular skills to serve his country, and then return to his life back home. He also makes a practice of discovering old trinkets in shop windows, and returns them to where they belong.

Imagine if he wrote haiku.

They all looked alike.
All the dead had just one face;
it was very young.

I remember the heady days of getting ready for war. I remember having a sense of collective purpose, and feeling like history was being made around me every day. I remember being a small part of big things. I remember recognizing that life and death were then part of a daily equation, if not for me directly, then for people I knew and loved. I do not miss those days. And yet, at the same time, I also miss them terribly. Things seemed clearer then. Black and white.

Now, like Stovall in repose while contemplating a post-war meadow, I often find myself flashing back to these times, both good and bad. I recall friends lost in time and space. I see our young faces. I think about the lessons we may have learned. I wonder which lessons are the most essential—the ones I most need to share with my children.

This is the question that most-haunts me: Do I tell them that "maximum effort" is a tool, a crutch, or a stick?

What did you set out to do? How did things turn out? What would you do differently, if you did it all again?

Epilogue

The summer grasses,
all that remains
of soldiers' dreams.

— Mastsuo Bashō (1644-1694)
translated by Lucien Stryk[1]

---

[1] Basho, Matsuo and Lucien Stryk. *On Love and Barley: Haiku of Basho*. English. Selections. London: Penguin Books, 1985. Print. Poem No. 252, p. 80

Other Poems, Other Lessons

10 haiku about Operation Desert Storm

1.
Boonie hat. Suede boots.
"Chocolate-chip" camouflage.
Gas mask on my hip.

2.
Origami plane /
a stealthy crease in the dark /
radar doesn't see.

3.
Saddam promises
the "Mother of All Battles"
if we take Kuwait.

4.
Anti-aircraft guns
shooting stars over Baghdad.
Show made for T.V.

5.
Voice of Darth Vader
adds commercial gravitas:
"*This* ... is C.N.N."

6.
Laser-guided gaze
reveals war is really just
Super Nintendo.

7.
The nightly sports scores:
"Scuds vs. the Patriots"—
bombs bursting in air!

8.
Midnight at high noon.
Oil wells burn the air black.
War is hell on Earth.

9.
A "Line in the Sand."
A hundred-hour ground war.
A "Highway of Death."

10.
Pottery Barn rule
said "You break it, you own it."
That's why we're still there.

sound of a left-handed baseball bat, clapping

*"embrace the
Suck."*

*"if you don't mind,
it don't matter."*

*"if it ain't raining,
it ain't training."*

now, I wonder at how
it took me years to figure out:

my drill sergeant
was, in reality, a Buddhist monk.

toward a poetics of lessons-learned

A "lesson" is: knowledge
gained from experience.

A "lesson-learned" is: knowledge
gained from experience

applied to change
individual or group
behavior.

A lesson-learned is "integrated"
when it is shared
with others.

From this ground,
five corollaries grow:

1.
There are no mistakes
except for ignoring results.

2.
Second chances count.
Often, more than firsts.

3.
Risks can be mitigated,
not eliminated.

4.
The safety gods may be appeased
only temporarily. They routinely demand sacrifice.

5.
In war, doing everything right
can still get you killed.

Try not to learn
that last one
the hard way.

Kintsugi

Ashikaga Yoshimasa sent the shards away with hope
that artisans could somehow fashion
a repair for his shattered bowl.

Lacquered gold now fills its cracks;
it is stronger in the broken places.

The helmet that saved the life of Army Specialist Tom Albers
was shipped off to the procurement program executive office.
After months of analysis, it was eventually returned

to sit in a trophy case.

Better Hooches and Gardens

The Old Man sat with me
on the back end of an Army truck
waiting to go into The Box.

Told me he always knew
I'd somehow make a magazine spread out of this.
Maybe because making a war is like remodeling a house:

Demolition is the easy part,
every project costs more than it should,
and even the glossy before-and-after pictures

tell doubtful truths.

RANDY BROWN

a quiet professional professes in haiku

1.
A practice of war
involves daily sacrifice.
The job is a trade.

2.
This we will defend:
Constitution, people, land.
(The order matters.)

3.
Any rag-bag Joe
who ever raised their right hand?
Now also, my kin.

4.
The only glory
one should seek is the respect
of one's own soldiers.

5.
"Secret" means secret.
Loose lips sink ships, lives, careers.
Keep your big trap shut.

6.
Your moral compass
should be red-light readable
for work in the dark.

7.
Share knowledge freely.
A lesson-learned is like cheap
immortality.

Toward an understanding of war and poetry, told (mostly) in aphorisms

Poetry is the long war of narrative.

Poetry, like history, is subjective.

If journalism is the first draft of history, poetry is the last scrap.

Poets set the stage of victory. Just ask Homer: *Who won the ball game?*

Do not make fun of war poets. A war poet will cut you.

War is hell. Poetry is easier to read. But each takes time.

Any war poem is a final message home.

Poetry can survive fragmentation. Irradiation. Ignorance.

Poetry can cheat death.

Poetry has all the time in the world.

Poetry will outlast us all.

Poetry is a cockroach.

"History does not repeat itself, but it does rhyme."—Mark Twain

"Twain didn't actually say that."—John Robert Colombo[1]

John Robert Colombo is a poet.

---

[1] While John Robert Colombo incorporated the popular "history rhymes" quotation—which he then attributed to Mark Twain—into his 1970 work, "A Said Poem," he later privately reported he was uncertain of its origins. (Full disclosure, however: Despite the poetic construction here, Colombo himself never said, "Twain didn't actually say that.") In an 1874 introduction to *The Gilded Age: A Tale of To-Day*, a novel co-written with Charles Dudley Warner, Twain apparently *did* say, "History never repeats itself, but the Kaleidoscopic combinations of the pictured present often seem to be constructed out of the broken fragments of antique legends." History prefers Colombo's version. So do I.

an Army lessons-learned analyst writes haiku

1.
A "lesson-learned" is
knowledge from experience.
Share yours and be wise.

2.
Make your decisions
faster than the other guy's:
Eat his OODA[1] loops.

3.
Any change can be
ascribed to seven factors
called "dot-mil-pee-eff."[2]

4.
"Make no small plans," but
"the enemy gets a vote"—
what *can* go wrong *will*.

5.
Review your results:
Candid discussions should make
better luck next time.

---

[1] An "OODA loop" is the decision cycle of "observe," "orient," "decide," and "act," developed by military strategist and U.S. Air Force fighter pilot Col. John Boyd.
[2] The mnemonic "DOTMLPF" stands for "Doctrine," "Organization," "Training," "Materiel," "Leadership," "Personnel," and "Facilities"—all potential factors in creating or analyzing organizational change.

Humility

In high school, before the Internet, Miss Barbara Hess
dispatches students of World War II to the library,
    arming each with a name:

"Research your topic. Write me a letter as that person. You will know
what I want you to learn, when you have found it."

Richard O'Kane, commander, USS Tang, on his fifth submarine patrol
October 1944 in the Formosa Strait, engages a target. Fires two fish.

The Tang's last torpedo porpoises, circles back, and strikes its own boat.
78 sailors are killed. When Kane and 8 others get to the surface,
    they are beaten by their Japanese rescuers.

In 2020, an acting Secretary of the Navy travels across the Pacific
to personally address the crew of USS Theodore Roosevelt.

Off-the-cuff, over the ship's public address system,
    the would-be Man in the Arena
calls their recently relieved-but-beloved captain either "too naive
    or too stupid."

The torpedo circles back.

contact print

In Africa, after a Stuka had strafed overhead, reporter Ernie Pyle tapped
the shoulder of the dead soldier
next to him and asked,
"Whew, that was close, eh?"

On Ie Shima, 23 square kilometers of Indiana-flat
    southwest of Okinawa,
Ernie Pyle ditched out of a Jeep,
but then caught a burst of .30-caliber machine-gun fire
under his helmet, over the eyebrow, left temple.

There was a photo, unreleased—no negatives remain—
captured by Alexander Roberts on a Speed Graphic
at the end of a laborious, dirt-eating crawl to the body.
"It was so peaceful a death ..." he said 63 years later,

"that I felt its reproduction would not be in bad taste."

Icarus over Wisconsin

The twin turbocharged Allison V-12 engines churned the air
    in a butterfly stroke,
the props pulling pilot Richard "Dick" Bong, 23, of nearby Poplar,
in a silver-winged gondola, slipping above the shipyards of Superior,
    Wisconsin.

By the end of the war, driving a forked-tail Lightning named "Marge,"
with an icon of his wife on the side of his plane,
    Bong would be responsible
for downing a total of 40 sons of the Rising Sun. "Flew straight at
    them," he said.

Hailing Bong as "Ace of Aces," Douglas McArthur himself
pinned on the nation's highest medal and welcomed
    the former farm boy
to the halls of the bravest of the brave.

In late '45, victim of an unflipped switch and an aux water pump,
the *L.A. Times* announced the new test pilot's death in a jet,
the impact heralded on the front page, in larger type,
    and above the mention

that we'd gone ahead and dropped the Bomb.

RANDY BROWN

Clausewitzian nature poem

the only thing
war ever changes
is the uniform

Secondary Targets

Recommended Books, Movies & Poems

I offer these additional notes for those who might wish to dive more deeply into topics such as U.S. Air Force history during World War II; United States air power doctrine, and the real-life intersections of military culture and the "Twelve O'Clock High" franchise.

In other words, if you like movies and TV shows and poems and books about B-17 bombers, *bombs away!*

❖ ❖ ❖

The Eighth Air Force is an overwhelmingly popular topic on which to publish memoirs, movies, and fact-books. Here are a few I found useful, particularly in gaining an appreciation of life as a U.S. bomber air crew member in World War II England:

Donald L. Miller's 2006 non-fiction book *Masters of the Air: America's Bomber Boys Who Fought the Air War Against Nazi Germany* synthesizes many of the organizational, psychological, and physical hazards navigated in the creation of the early history of that organization. It is the foundation of a forthcoming mini-series to be distributed via the Apple TV+ digital streaming service. Imagine "Band of Brothers" (2001) or "The Pacific" (2010), but with bombardiers. Gerald Astor's *The Mighty Eighth: The Air War in Europe as Told by the Men Who Flew It* is a popular and accessible oral history collection, upon which readers can assess for themselves fictional World War II-stories against the voices of those who experienced the real thing.

To those readers who wish to consider strategic bombing from a higher altitude, I recommend R.J. Overy's *The Bombers and the Bombed: Allied Air War Over Europe, 1940-1945*. Regarding the Pacific Theater, I suggest Kenneth P. Werrell's *Blankets of Fire: U.S. Bombers over Japan during World War II*; and Barrett Tillman's *Whirlwind: The Air War Against Japan, 1942-1945*.

❖ ❖ ❖

Not surprisingly, both fiction and non-fiction films figure as primary sources in any consideration of the "Twelve O'Clock High" franchise. Some of these, including "Target for Tonight" (1941) and "Target for Today" (1944), were military-produced documentaries from which footage might have been repurposed for "Twelve O'Clock High." The former follows a full mission cycle—beginning, middle, and end—of a British bomber group's nighttime operation. The latter follows the American daytime equivalent. Both are readily available via Internet and streaming services.

The 45-minute color documentary "Memphis Belle: The Story of a Flying Fortress" (1944) focuses on a single mission of a B-17 bomber crew. Director William "Willy" Wyler was already well-established as a civilian film-maker when he was recruited in part by U.S. Army Capt. Sidney "Sy" Bartlett.[1] Bartlett, the future co-writer of "Twelve O'Clock High," was at the time acting as something as a headhunter of Hollywood movie makers[2]. Bartlett's future "Twelve O'Clock High" writing partner, Lt. Col. Beirne Lay Jr., was later instrumental in getting Maj. Willy Wyler attached to the Eighth Air Force. There, he arranged for Wyler and his camera operators to receive flight training, that they might film aboard B-17 bombers on combat missions.[3] While Wyler was already working overseas in uniform on the "Memphis Belle" project, his pre-war, pro-British movie "Mrs. Miniver" (1942) won six Academy Awards—including best director and best picture.[4] (Of course, the movie can be considered "pre-war" with regard only to the U.S. market—at the time of its release, British movie houses were already under German bomber attack.) When Wyler's B-17 documentary was theatrically released to U.S. audiences in 1944, it became the first movie in history to be reviewed on the front page of *The New York Times*.[5] The

---

[1] Harris, Mark. *Five Came Back: A Story of Hollywood and the Second World War*. New York: Penguin Books, 2015. Print. pp. 123, 139
[2] Harris, p. 111
[3] Harris, pp. 200-201
[4] Harris, p. 205
[5] Harris, p. 297

documentary is readily viewable via Internet and streaming services.

Raw color-film footage originally shot for Wyler's original World War II "Memphis Belle" was later rediscovered, restored, and repurposed for another documentary, "The Cold Blue" (2018). For a more extensive exploration of Wyler's wartime movie-making, seek out the Netflix documentary mini-series "Five Came Back," as well as the Mark Harris book of the same name.

The feature film "Memphis Belle" (1990), starring Matthew Modine, Eric Stoltz, D.B. Sweeney, and many other fine actors, was co-produced by Wyler's daughter Catherine. Often mis-labelled as a "remake" of the 1944 documentary, the fictional story seems to dramatically incorporate many of experiences and anecdotes taken from Eighth Air Force histories.

While there are many published and archived collections of Eighth Air Force oral history, Suzanne Broderick's *Real War vs. Reel War: Veterans, Hollywood, and WWII* specifically interrogates B-17 pilot Jim Oberman's wartime recollections with his impressions of "Memphis Belle" (1990) and "Twelve O'Clock High" (1949).[6]

Regarding other feature films featuring stories similar or related to "Twelve O'Clock High," the contemporaneous "Command Decision" (1948) is particularly worthy of note. Unlike "Twelve O'Clock High," it centers on the frictions among officers and politicians, as each stakeholder attempts to advance their respective careers, positions, and philosophies regarding daylight precision bombing. The movie is based on a 1948 play and novel written by William Wister Haines, who had served in England as a U.S. Army Air Forces intelligence officer.

In "Command Decision," actor Chuck Gable plays a U.S. bomb group commander. Gable was himself a veteran of the Eighth Air Force,

---

[6] Broderick, Suzanne "Stormy Weather: Memphis Belle and Twelve O'Clock High" *Real War vs. Reel War: Veterans, Hollywood, and WWII*. Rowman & Littlefield, 2015. Broderick reports that Oberman, a 35-mission veteran of the Eighth Air Force, enthusiastically endorses the "Memphis Belle" documentary, thinks the 1990 feature was too dramatic, and effusively praises the realism of "Twelve O'Clock High"—"down to the minutest detail." Print. p. 52

having served on at least five missions as a B-17 waist gunner. During his time in England, he also directed and produced the 62-minute color documentary "Combat America" (1943). Originally conceived as a recruiting product, the film was released more as an historical snapshot of Eighth Air Force life and operations. The chummy footage features a lot of good-natured (if sometimes a bit scripted) ribbing among Gable and his fellow airmen.

❖ ❖ ❖

In *The World War II Combat Film: Anatomy of a Genre*, Jeanine Basinger carefully defines what a "combat movie" is and is not, and how the genre has evolved over time. In Basinger's "Second Wave"—films made following U.S. entry into the war—both directors and viewers had to first learn the tropes, both in terms of expected story points (example: "A hero has leadership forced upon him in dire circumstances"[7]) and visual cues (example: "During opening credits, a message of dedication to the troops is displayed"[8]). In this period, "Documentaries taught viewers what real combat looked like, while filmmakers created formal, visual shorthand for narrative concepts the audience had learned."[9]

By the post-war period, movie-goers had learned what to expect from a war story on the silver screen. Films in Basinger's "Third Wave" (1949-1959) involve "creation of a filmed reality based on earlier films and history, with conscious use of genre."[10] With its re-use of World War II gun-camera footage, and adherence to expected genre trappings, "Twelve O'Clock High" seems to land squarely on being one of the first of this "Third Wave."

Basinger's extensive taxonomy is not only useful in dissecting war movies, but in discovering other projects that may connect to the

---

[7] Basinger, Jeanine. *The World War II Combat Film: Anatomy of a Genre*. Middletown, Conn: Wesleyan University Press, 2003. Print. p. 68
[8] Basinger, p. 67
[9] Basinger, p. 112
[10] Basinger, p. 140

"Twelve O'Clock High" franchise in theme, story, and/or production. Take, for example, the 1943 movie "Bombardier" (1943). Although it is 90 percent a "training camp" narrative, Basinger writes, she seems to suggest it still makes her cut as a "combat film," because the story contains "an Armageddon of actual combat as a finale"[11] I recommend the movie to "Twelve O'Clock High" buffs. Despite its reliance on model effects—and its boisterous boosting of bombardiers at the expense of other crew specialties (there is even a come-join-the-bombardiers sing-along)—the movie depicts some of the aircraft, analog gadgets, and techniques used to train B-17 bomber crews for daylight precision bombing. More subtly, the movie also briefly touches upon contemporary debates regarding developing doctrine, organization, training, and personnel.

(Less enthusiastically, I note also "Air Force," a 1943 movie that, as Steve Call mentions in *Selling Air Power*, "not only depicts the exploits of a B-17 crew in the opening days of the war, it elevates the technical superiority and bombing accuracy to absurd levels."[12])

In Basinger's "Fifth Wave" of combat films (roughly 1965-1975), combat films begin to subvert genre forms by presenting "inverted, parodied, satirical, and opposite" stories of war. It is fascinating to consider that "12 O'Clock High" existed on late-1960s television in roughly the same airspace as Stanley Kubrick's dark Cold War comedy "Dr. Strangelove, Or: How I Learned to Stop Worrying and Love the Bomb" (1964) and the adaptation of Joseph Heller's World War II satirical novel, "Catch-22" (1970). Applying Bassinger's language to television: "12 O'Clock High" was a "Third Wave"-style narrative launched during "Fifth Wave" times.

❖ ❖ ❖

---

[11] Basinger, p. 112
[12] Call, Steve. *Selling Air Power: Military Aviation and Popular Culture after World War II*. College Station, Texas, Texas A&M University Press, 2009. Print. p. 27

I wanted my pithy poems about air power to be well-grounded in history. Specifically, I wanted to make sure I understood how a nation's military comes to develop a war-fighting concept in thought and action. For most armchair historians, Donald L. Miller's *Masters of the Air* will likely prove sufficient background for a fuller appreciation of "Twelve O'Clock High." For truly dedicated students of air power, however, there is a potential onslaught of weighty scholarship available.

Tami Davis Biddle's *Rhetoric and Reality in Air Warfare: The Evolution of British and American Ideas about Strategic Bombing, 1914-1945* is a pathfinder tome for other air power historians, from which many other scholastic sorties might be launched. Stephen L. McFarland's *America's Pursuit of Precision Bombing, 1910-1945*, for example, focuses on the development of daylight precision bombing through a lens of engineering and acquisition. I found these respective explorations of rhetoric and tech enriched my understanding of daylight precision bombing.

For purposes of brevity in *Twelve O'Clock Haiku*, I admittedly leveraged 1941's U.S. Air War Plans Division plan No. 1 (AWPD-1) as something of a doctrinal stone tablet—an object that seems to have suddenly occurred, as if delivered from deities on high. The pre-war evolution of U.S. air power doctrine is more expertly and comprehensively covered in Craig Morris' *The Origins of American Strategic Bombing Theory*.

I also found it useful to explore this period of doctrinal development by way of biography. Here, I recommend Charles Griffith's *The Quest: Haywood Hansell and American Strategic Bombing in World War II*. This useful title that is available for free via the U.S. Air Force's Air University, both as a print book and as a digital download.[13] I also found Brian Laslie's *Architect of Air Power: General Laurence S. Kuter and the birth of the U.S. Air Force* a particularly useful tonic against some of the

---

[13] Griffith, Charles. *The Quest: Haywood Hansell and American Strategic Bombing in World War II*. Maxwell Air Force Base, Alabama: Air University Press, 1999. www.airuniversity.af.edu/Portals/10/AUPress/Books/B_0073_GRIFFITH_QUEST_HAYWOOD_HANSELL.pdf

breathless hagiography surrounding the mythic "bomber mafia" pantheon.

Steve Call's *Selling Air Power: Military Aviation and Popular Culture after World War II* helped me connect the air power doctrine discussion to consideration of popular films such as "Twelve O'Clock High." As someone who likes watching how information flows and changes within an organization, I also appreciated the "marketing" framing of Brian D. Vlaun's *Selling Schweinfurt: Targeting, Assessment, and Marketing the Air Campaign Against German Industry*.

For good or ill, Malcom Gladwell's podcast-generated *Bomber Mafia* was published during the writing of *Twelve O'Clock Haiku*. While often criticized for errors in fact and over-simplification, Gladwell spins a good story, particularly in positioning U.S. air generals Haywood Hansell and Curtis LeMay as opposing intellectual and moral archetypes.

I would, however, encourage readers to explore the history more widely, in order to assess for themselves some of Gladwell's storytelling choices. For example, I found Daniel T. Schwabe's *Burning Japan: Air Force Bombing Strategy Change in the Pacific* an essential and readable history describing the step-by-step evolution of organizational decision-making between the two generals' respective commands. Also, the first two multi-chapter sections of James M. Scott's *Black Snow: Curtis LeMay, the Firebombing of Tokyo, and the Road to the Atomic Bomb* engagingly illuminate how Hansell's intellectual and moral commitments to precision-bombing doctrine were challenged by a chorus of subordinates, peers, and superiors.

One might say that LeMay didn't kill daytime precision bombing; he only pulled the trigger.

❖   ❖   ❖

Finally, a quick salvo for those who might wish to further consider the interconnections among poetry and aerial warfare in World War II. In *Words to Measure a War*, David K. Vaughn analyzes the poetry and biographies of nine veterans of World War II, five of whom are veterans of the U.S. Army Air Forces. In alphabetical order, these are:

John Ciardi (1916-1986), who served as an aerial gunner on B-29 "Superfortress" heavy bombers in the Pacific Theater.

James Dickey (1923-1997), who served in the Pacific as a radar-operator on a P-61 "Black Widow" all-weather fighter. From 1966-1968, he was the poetry consultant to the U.S. Library of Congress—a position now known as the U.S. poet laureate.

Richard Hugo (1923-1982), who flew 35 missions as a bombardier on a four-engined B-24 "Liberator" bomber. Assigned to a base in Italy, he flew missions against targets in Southeastern and Eastern Europe as part of the Fifteenth Air Force.

Randall Jarrell (1914-1965), who scrubbed out of military pilot training, but became instead a stateside instructor. He reportedly considered his job title—"celestial navigation tower operator"—the most poetic in the Air Force.[14] Jarrell held the equivalent position to U.S. poet laureate 1956-58.

Howard Nemerov (1920-1991), who joined the Canadian armed forces in order to become a pilot, and later transferred into the U.S. Army Air Forces. He flew anti-shipping missions in a twin-engine fighter-bomber called a Bristol "Beaufighter." Nemerov was the U.S. poet laureate from 1963-64 and again in 1988-1990. His collection *War Stories: Poems about Long Ago* (1987), contains a 14-poem section titled "The War in the Air."

Given the American-heavy-bombers-over-Europe flavor of *Twelve O'Clock Haiku*, I particularly commend to readers the works of Randall Jarrell. Jarrell's stateside job placed him in a privileged position from which to observe and overhear air crew stories, which he often channeled into poetic personas. His "The Death of the Ball Turret Gunner" (1945) is brutal and short, and much-anthologized in print and on-line. Regarding bombers, there is also his "Eighth Air Force" (1945) and "Second Air Force" (1945). (Second Air Force was the stateside training command tasked with generating pilots, navigators, bombardiers, and other U.S. air crew.) Given *Twelve O'Clock Haiku's* particular focus on "lessons-learned," I also wish to mention Jarrell's "The Learners" (1945).

---

[14] "Randall Jarrell, Poet, Killed by Car in Carolina" *New York Times*, Oct. 15, 1965

Then only 5 years old, American poet James Tate's father was a B-17 pilot who was killed on a bombing mission to Stuttgart. The rest of the crew survived. His father's remains were never recovered. Tate's "The Lost Pilot" served as the title poem to his 1967 print collection.

With similar motives, I also recommend Daniel Swift's literary exploration of his grandfather's experiences as an RAF heavy-bomber pilot. The creative non-fiction work is *Bomber County: The Poetry of a Lost Pilot's War*, and includes discussions of both allied and American poets. Of particular note is the equal consideration of poems regarding those who are doing the bombing, and those who are being bombed.

Found elsewhere, there is an apologetic 1977 poem-letter by U.S. bombardier-turned-poet Richard Hugo, which is addressed to Serbian-American poet (and eventual U.S. poet laureate) Charles Simic. In "Letter to Simic from Boulder,"[15] Hugo illuminates his likely role in bombing Simic when the latter was a youth growing up in the city of Belgrade, Yugoslavia. The poem was sparked by a chance 1972 meeting in San Francisco,[16] in which Hugo was able to sketch a map of Simic's hometown, despite never having visited. ("I only bombed it a few times."[10] Hugo off-handedly said at the time.) After seeing Hugo's distress at the realization that Simic was likely one of those bombed, Simic assured him there were no bad feelings.

Earlier, after twice returning in the 1960s to his former haunts in Italian fields, meadows, and cantinas, Hugo wrote a war-poetry collection titled *Good Luck in Cracked Italian*. "I still wasn't sure why I'd come back," he writes in an autobiographical essay titled "Ci Vediamo." (Translated, the title means "We will see each other") "But I felt it must be the best reason in the world."[17]

In the essay, Hugo recounts an epiphany he'd experienced during the

---

[15] Hugo, Richard. "Letter to Simic from Boulder" first published in *31 Letters and 13 Dreams: Poems*. Norton, 1977. Print.

[16] Simic, Charles. *A Fly in the Soup: Memoirs*. Ann Arbor: University of Michigan Press, 2005. Print. pp. 12-14

[17] Hugo, Richard, Ripley S. Hugo, Lois M. Welch, and James Welch. "Ci Vediamo," *The Real West Marginal Way: A Poet's Autobiography*. W.W. Norton, 1992. Print. p. 129

war. The bombardier had gotten lost while attempting to hitchhike back to his base, and sat down in a meadow that—for me, at least—evokes the fields of Elysium and Archbury:

"I was tired, dreamy, the way we get without enough sleep, and I watched the wind move in waves of light across the grass. The field slanted and the wind moved uphill across it, wave after wave. The music and motion hypnotized me. The longer the grasses moved, the more passive I became [...] I didn't care about getting back to the base now. I didn't care about the war. I was not part of it anymore [...]"[18]

The field is explicitly mentioned in two of Hugo's poems, each of which appears in the essay: "Centuries Near Spinnazola" and "Spinazzola: Quella Cantina Là."[19] Michael S. Allen calls the latter poem the most important of Hugo's career.[20] According to William Matthews, Hugo grew up admiring the masculinity of local tough guys in south side Seattle, Washington. Later, Hugo extended that admiration to "sardonic movie stars, detective heroes, and British Royal Air Force flyers."[21] The poem "Spinazzola: Quella Cantina Là" is key to Hugo, Allen writes, because it marks the poet's transcendence—from defining masculine strength in terms of movie-star bravado, to "the more-enduring toughness of the human heart."[22]

❖ ❖ ❖

The writing of this project has been a journey. As I mentioned in the introduction, *Twelve O'Clock Haiku* started as a whimsical experiment,

---

[18] Hugo, pp. 114
[19] Hugo, *The Real West Marginal Way*, Print. pp. 124-126. In the essay, Hugo addresses his misspelling of the place name "Spinazzola," which occurred in the original publication of the poem "Centuries Near Spinnazola."
[20] Allen, Michael S. *We Are Called Human: The Poetry of Richard Hugo*. Fayetteville: University of Arkansas Press, 1982. Print. pp. 72-73
[21] Matthews, William. "Introduction," *The Real West Marginal Way: A Poet's Autobiography*. Richard Hugo, Ripley S. Hugo, Lois M. Welch, and James Welch. W.W. Norton, 1992. Print. p. xvi
[22] Allen, pp. 72-73

turned into an obsession, and ended up generating a cascade of epiphanies.

Uncle Sam taught me that a "lesson-learned" is knowledge gained from experience; it is considered "integrated" when it is shared with others. For me, many lightning-bolts occurred late in the project's development. After immersing myself in film and TV criticism, World War II history, and air power doctrine, I read "bomber poetry" with new eyes. In my encounters with historical poets such as Santōka Taneda and Richard Hugo, I found new friends and fellow travelers. Whenever I worried that I was in danger of losing the bubble—that all this sharp and heavy research would ultimately puncture my original, light-hearted purpose—a few lines of poetry from the past would glint through the clouds, and point the way home.

War poetry is a thin red line, connecting us all, regardless of age, background, or era. My friend and fellow poet Abby E. Murray recently quipped via social media: *"Every poem is a how-to poem. Change my mind."* That made me laugh, but it also made me think. In such moments of laughter and poetry, the universe speaks to us. Every poem is a how-to poem. Every poem is a last letter home. Every poem is a potential lesson shared.

❖ ❖ ❖

In the poem "Spinazzola: Quella Cantina Là,"[23] Richard Hugo asks:

[...] Don't our real friends
tell us when we fail? Don't honest fields
reveal us in their winds? [...]

All we have to do is listen.

---

[23] Hugo, Richard. *Making Certain It Goes on: The Collected Poems of Richard Hugo.* Norton, 2007. Print. pp. 43-45

Bibliography

Allen, Michael S. *We Are Called Human: The Poetry of Richard Hugo*. Print. University of Arkansas Press, 1982. Print.

Aurelius, Marcus and George Long. *Meditations of Marcus Aurelius*, 2021. Print.

Astor, Gerald. The Mighty Eighth: The Air War in Europe As Told by the Men Who Flew It. New York, N.Y: D.I. Fine Books. Print. 1997

Baldwin, Stanley. "A Fear for the Future": Speech before the House of Commons of the United Kingdom, Nov. 10, 1932 Stanley Baldwin. On-Line: missilethreat.csis.org/wp-content/uploads/2020/09/A-Fear-for-the-Future.pdf

Bassinger, Jeanine. *The World War II Combat Film: Anatomy of a Genre*. Print. Wesleyan University Press, 2003

Barno, David and Nora Benashel "The Catastrophic Success of the U.S. Air Force." *War on the Rocks*, May 3, 2016. On-line: warontherocks.com/2016/05/the-catastrophic-success-of-the-u-s-air-force/

Basho, Matsuo and Lucien Stryk. *On Love and Barley: Haiku of Basho*. English. Print. London: Penguin Books, 1985

Biddle, Tami Davis. *Rhetoric and Reality in Air Warfare: The Evolution of British and American Ideas About Strategic Bombing, 1917-1945*. Print. Princeton University Press, 2002

Bodnar, John. *The "Good War" in American Memory*. Print. The Johns Hopkins University Press, 2010

Boehm, Richard Theodore. "Tod B. Galloway: Buckeye Jongleur,

Composer of 'The Whiffenpoof Song'" *Ohio History Journal* Vol. 83, No. 4, Autumn 1974, pp. 256-282.

"Bombardier." Film. Produced by Robert Fellows. Directed by Richard Wallace. Written by John Twise and Martin Rackin. Performances by Pat O'Brien, Randolph Scott, Anne Shirley, and Eddie Albert. 99 minutes. Black and white. 1943

Broderick, Suzanne. "Stormy Weather: Memphis Belle and Twelve O'Clock High." *Real War vs. Reel War: Veterans, Hollywood, and WWII*. Print. Rowman & Littlefield, 2015

Buchanan, Leigh and Mike Hofman. "Everything I Know about Leadership, I Learned From the Movies" *Inc. Magazine,* March 1, 2000. On-Line: www.inc.com/magazine/20000301/17290.html

Call, Steve. *Selling Air Power: Military Aviation and Popular Culture after World War II*. Print. Texas A&M University Press, 2009

Clodfelter, Mark. *Beneficial Bombing: The Progressive Foundations of American Air Power, 1917-1945*. Print. Bison Books, 2013

Coffey, Thomas M. *Iron Eagle: The Turbulent Life of General Curtis LeMay*. Crown Publishers, Inc., 1986

"Command Decision." Film. Directed by Sam Wood. Screenplay by William R. Laidlaw and George Froeschel. Original play and novel by William Wister Haines. Performances by Clark Gable, Walter Pidgeon, Brian Donlevy, Van Johnson. 112 minutes. Black and white. 1948

"Combat America." Film. Written by John Lee Mahin. Narrated by and featuring Clark Gable. First Motion Picture Unit, U.S. Eighth Air Force. 62 minutes. Color. 1945

Correll, John. T. "Daylight Precision Bombing" *Air Force Magazine,*

Oct. 8, 2008. On-line: www.airforcemag.com/article/1008daylight/
—. "The Real Twelve O'Clock High." Jan. 1, 2011 *Air Force Magazine*. On-line: www.airforcemag.com/article/0111high/

Davis, William V. "'Good Luck in Cracked Italian': Richard Hugo in Italy" *War, Literature & the Arts Journal: An International Journal of the Humanities*. 2008. Vols. 1 and 2, pages 57-73. On-line: www.wlajournal.com/wlaarchive/20_1-2/Davis.pdf

Dickey, Colin "Malcolm Gladwell's Fantasy of War From the Air" (Review of Gladwell's *Bomber Mafia*). *New Republic*, June 4, 2021. On-line: newrepublic.com/article/162624/malcolm-gladwells-fantasy-war-air-bomber-mafia-review

Doherty, Thomas Patrick *Projections of War: Hollywood, American Culture, and World War II* Columbia University Press, New York, New York 1999

Duffin, Allan T. and Paul Matheis. *The 12 O'clock High Logbook: The Unofficial History of the Novel, Motion Picture, and TV Series*. Kindle ed. BearManor Media, 2005

D'Zurilla, Christie. "Ryan Reynolds explains what his 'little sabbatical' from making films is really about." *Los Angeles Times*, Nov. 3, 2021. On-line: www.latimes.com/entertainment-arts/movies/story/2021-11-03/ryan-reynolds-acting-movies-sabbatical

Edwards, Sam. "12 O'Clock High and the Image of American Air Power, 1946-1967." *American Militarism on the Small Screen*, edited by Anna Froula and Stacy Takacs. Print. Routledge, 2016

Erickson, Glenn. "Twelve O'Clock High – Fox Reissue" *DVD Savant*, 2007. On-line: www.dvdtalk.com/dvdsavant/s2324high.html

Etter, Jonathan. *Quinn Martin, Producer: A Behind the Scenes History of QM Productions and Its Founder*. Print. McFarland, 2004

"Five Came Back." Three-episode documentary mini-series via Netflix. Directed by Laurent Bouzereau. Written by Mark Harris. Performances by Francis Ford Coppola, Guillermo del Toro, Paul Greengrass, Lawrence Kasdan, Steven Spielberg. Narrated by Meryl Streep. 2015

Fleischman, John. "Some of Us Have Got to Die: What the 1949 film Twelve O'Clock High still tells us about air combat and the burden of command." *Air & Space Magazine,* October 2018. On-line: www.airspacemag.com/history-of-flight/twelve-oclock-high-180970369/

Freeman, Roger A. *The Mighty Eighth: Units, Men and Machines*. Print. Doubleday, 1970
—. *The Mighty Eighth War Manual*. Print. Cassell, 2001

Froula, Anna and Stacy Takacs. *American Militarism on the Small Screen*. Routledge, 2016

Fussell, Paul. *Wartime: Understanding and Behavior in the Second World War*. Print. Oxford University Press, 1989
—. *Thank God for the Atom Bomb, and Other Essays*. Print. Ballantine Books, 1990

Gerstenberger, Donna L. *Richard Hugo*. Print. Boise State University Western Writers Series No. 59. Boise State University, 1983

Gravatt, Brent L., and Francis H. Ayers. "The Fireman: Twelve O'Clock High Revisited." *Aerospace Historian*, vol. 35, no. 3, pp. 204–08, 1988

Griffith, Charles. *The Quest: Haywood Hansell and American Strategic Bombing in World War II*. Print. Air University Press, 1999

Hansell Jr., Haywood S. *The Air Plan that Defeated Hitler*. Print. Higgins-McArthur/Longino & Porter, 1972

Harris, Mark. *Five Came Back: A Story of Hollywood and the Second World War*. Print. Penguin Books, 2015

Headquarters, Department U.S. Army (2017). "The Army Lessons Learned Process" (Army Regulation 11-33). On-line: armypubs.army.mil/epubs/DR_pubs/DR_a/pdf/web/ARN2887_AR11-33_Web_FINAL.pdf

Headquarters, Department U.S. Army (2019). "Army Leadership and the Profession" (Army Doctrine Publication 6-22). On-line: armypubs.army.mil/epubs/DR_pubs/DR_a/ARN20039-ADP_6-22-001-WEB-0.pdf

Hugo, Richard. "*Catch 22*, Addendum" and "Ci Vediamo" *The Real West Marginal Way: A Poet's Autobiography*. Ripley S. Hugo, Lois M. Welch, and James Welch, editors. W.W. Norton, 1986
—. *Making Certain It Goes On: The Collected Poems of Richard Hugo*, W.W. Norton, 1984

Hurley, Alfred F. and Robert C. Ehrhart (eds.) "Air Power and Warfare: The Proceedings of the 8th Military History Symposium" Office of U.S. Air Force History and U.S. Air Force Academy, Washington Oct. 18-20, 1978

Insinga, Richard. *Participant Guide for Twelve O'Clock High*. Print. Hartwick Classic Film Leadership Cases, The Hartwick Humanities in Management Institute. Hartwick College, Oneonta, New York, 2001
—. *Teaching Notes for Twelve O'Clock High*. Print. Hartwick Classic Leadership Case, The Hartwick Humanities in Management Institute. Hartwick College, Oneonta, New York, 2001

Jarrell, Randall. *The Complete Poems*. Farrar, Straus, and Giroux, 1969

Laslie, Brian D. *Architect of Air Power: General Laurence S. Kuter and the birth of the U.S. Air Force*, Kindle ed. The University Press of Kentucky, 2017

Lay Jr., Beirne and Sy Bartlett. Print. *Twelve O'Clock High!* Bantam Books, 1949

Lay Jr., Lt. Col Beirne. "I Saw Regensburg Destroyed" *Saturday Evening Post*, Nov. 6, 1943
—. "What It Takes to Bomb Germany: The Kind of Men Who Can Do It" *Harper's magazine*, Vol. 187, No. 1122. November 1943

Liu, Chen-ou. "Poetic Musings: Summer Grass Haiku by Basho" NeverEnding Story blog. June 10, 2013. On-line: neverendingstoryhaikutanka.blogspot.com/2013/06/poetic-musings-summer-grass-haiku-by.html

McFarland, Stephen L. *America's Pursuit of Precision Bombing, 1910-1945*. Print. University of Alabama Press, 2008

"Memphis Belle: A Story of a Flying Fortress." Film. Directed by William Wyler. Screenplay by Jermome Chodorov, Lester Koenig, William Wyler. Narrated by Lester Koenig. 43 minutes. Color. 1941

"Memphis Belle." Film. Directed by Micael Caton-Jones. Screenplay by Monte Merrick. Performances by Matthew Modine, Eric Stolz, Tate Donovan, D.B. Sweeney, Billy Zane, Sean Astin. Produced by David Puttnam and Catherine Wyler. 107 minutes. Color. 1990

Miller, Donald L. *Masters of the Air: America's Bomber Boys Who Fought the Air War Against Nazi Germany*. Kindle ed. Simon and Schuster, 2006

Morris, Craig F. *The Origins of American Strategic Bombing Theory*. Print. Annapolis Naval Institute Press, 2017

Nemerov, Alexander "Modeling My Father" *The American Scholar* Vol. 62, No. 4 (Autumn 1993), pp. 551-561

Nemerov, Howard. *War Stories: Poems about Long Ago and Now.* Print. University of Chicago Press,1989

O'Mara, Raymond P. *Rise of the War Machines: The Birth of Precision Bombing in World War II.* Print. Naval Institute Press, 2022

Overy, Richard J. *The Bombers and the Bombed: Allied Air War Over Europe, 1940-1945.* Print. Viking, 2014

Owen, Wilfred "Dulce et Decorum Est." Poetry Foundation. On-line: www.poetryfoundation.org/poems/46560/dulce-et-decorum-est

Randall Jarrell 1914-1965. Robert Lowell, Peter Taylor, and Robert Penn Warren, eds. Farrar, Straus & Girouix, 1967

"Randall Jarrell, Poet, Killed by Car in Carolina" *New York Times*, Oct. 15, 1965. On-line: archive.nytimes.com/www.nytimes.com/books/99/08/01/specials/jarrell-obit.html

Ricks, Thomas E. "A New Best Defense Contest: Okay, What Should a Military 'Professional' Profess?" *Foreign Policy*, Jan. 9, 2017. On-line: foreignpolicy.com/2017/01/09/a-new-best-defense-contest-okay-what-should-a-military-professional-profess/
—. "What is a military professional? Here's an answer in 90 words of haiku poetry" *Foreign Policy*. Jan. 24, 2017. On-line:
—. "OK, just what does a military professional profess (Best Defense contest Round III)", March 3, 2017. On-line: foreignpolicy.com/2017/03/03/ok-just-what-does-a-military-professional-profess-best-defense-contest-round-iii/

Rhodes, Richard. "The General and World War III." *The New Yorker*. June 19, 1995

Rubin, Steven Jay. "Chapter 3, Twelve O'clock High." *Combat Films: American Realism, 1945–2010*. Print. McFarland, 1981

Scannell, Vernon. *Not Without Glory: The Poets of the Second World War*. Print. Florence: Taylor and Francis, 1976

Schwabe, Daniel T. *Burning Japan: Air Force Bombing Strategy Change in the Pacific*. Print. Potomac Books, 2015.

Scott, James M. *Black Snow : Curtis Lemay the Firebombing of Tokyo and the Road to the Atomic Bomb*. Kindle ed. W. W. Norton & Company, 2022

Shindler, Colin. *Hollywood Goes to War: Films and American Society 1939-1953*. Print. Routledge, 2013

Simic, Charles. *A Fly in the Soup: Memoirs*. Print. University of Michigan Press, 2000

Taneda, Santōka, and John Stevens. *Mountain Tasting: Zen Haiku*. Print. Weatherhill, 1991

"Target for Tonight." Film. Directed by Harry Watt. Royal Air Force documentary. 48 minutes. Black and white. 1941

"Target for Today." Film. Directed by William Keighley. First Motion Picture Unit, U.S. Army Air Forces documentary. 90 minutes. Black and white. 1944

Tate, James. "The Lost Pilot." 1982. Poetry Foundation. On-Line: www.poetryfoundation.org/poems/47810/the-lost-pilot

Tennyson, Alfred Lord. "The Charge of the Light Brigade." 1854. Poetry Foundation. On-line: www.poetryfoundation.org/poems/45319/the-charge-of-the-light-brigade

Tillman, Barrett. *Whirlwind: The Air War against Japan, 1942-1945.* Print. Simon & Schuster, 2011

Tirpak, John A. "John T. Correll, Former Air Force Magazine Editor-in-Chief, Dies at 81" *Air Force Magazine* April 5, 2021. On-line: www.airforcemag.com/john-t-correll-former-air-force-magazine-editor-in-chief-dies-at-81/

"Twelve O'Clock High." Film. Directed by Henry King. Screenplay by Sy Bartlett and Beirne Lay Jr., based on their 1948 novel. Performances by Gregory Peck, Hugh Marlowe, Gary Merrill, Dean Jagger, Millard Mitchell 20th Century Fox. Produced by Daryl F. Zanuck. 132 minutes. Black and white. 1949

"12 O'Clock High." TV show. Produced by Quinn Martin. Performances by Robert Lansing, Frank Overton, John Larkin, and Paul Burke. Originally aired on American Broadcasting Company (ABC) television network. Black and white (Seasons 1 and 2); color (Season 3). 1964-1968

The United States Bombing Surveys (PDF). Air University Press, 1987. On-line: www.airuniversity.af.edu/Portals/10/AUPress/Books/B_0020_SPANGRUD_STRATEGIC_BOMBING_SURVEYS.pdf

U.S. Department of Defense, Chairman of the Joint Chiefs of Staff Instruction (CJCSI) 3010.02E dated 17 Aug. 17, 2016 "Guidance for developing and Implementing Joint Concepts." Sections A-3 to A-5

Van Riper, A. Bowdoin. "'Baa Baa Black Sheep' and the Last Stand of the

WWII Drama." *American Militarism on the Small Screen*, edited by Anna Froula and Stacy Takacs. Print. Routledge, 2016

Vaughan, David K. *Words to Measure a War: Nine American Poets of World War II*. McFarland, 2009

Vlaun, Brian D. *Selling Schweinfurt: Targeting, Assessment, and Marketing the Air Campaign Against German Industry*. Kindle ed. Naval Institute Press, 2020

Vardamis, Alex A. "Randall Jarrell's Poetry of Aerial Warfare" published in *War, Literature & the Arts Journal,* Spring 1990 Vol. 2, No. 1. On-line: www.wlajournal.com/wlaarchive/2_1/AlexAVardamis.pdf

Werrell, Kenneth P. *Blankets of Fire: U. S. Bombers Over Japan During World War II*. Print. Smithsonian Institution Press, 1996

Wolk, Herman S. "Decision at Casablanca" *Air Force Magazine*, Jan. 1, 2003. On-line: www.airforcemag.com/article/0103casa/

Wraith, David. "Is Twelve O'Clock High the definitive leadership movie?" *Movie Leadership*, May 29, 2013. On-line: www.movieleadership.com/2013/05/29/is-twelve-oclock-high-the-definitive-movie-about-leadership/

Editorial Style Notes

To help differentiate among "Twelve O'Clock High" in its various forms—popular novel, feature film, and TV program—I have adopted the following style rules throughout this book:
- *Twelve O'Clock High* refers to the 1948 novel.
- "Twelve O'Clock High" refers to the 1949 feature film.
- "12 O'Clock High" refers to the TV series, which originally aired 1964-1967.

❖ ❖ ❖

The U.S. Army Air Corps was formed July 2, 1926. Its successor, the U.S. Army Air Forces (USAAF), was formed June 20, 1941. Unless otherwise specified, all references in this book to air force personnel and resources regard this latter nomenclature and era.

Today's U.S. Air Force celebrates Sept. 18, 1947 as its birthday.

❖ ❖ ❖

The first names of fictional characters Lt. Gen. Patrick Pritchard and Maj. Donald "Doc" Kaiser are variously reported—or sometimes, omitted completely. I have opted to use Allan T. Duffin and Paul Matheis' *The 12 O'Clock High Logbook: The Unofficial History of the Novel, Motion Picture, and TV Series* as the definitive reference.

❖ ❖ ❖

The U.S. VIII Bomber Command was established Jan. 19, 1942. On Feb. 22, 1944 it was redesignated as the U.S. Eighth Air Force, now sometimes popularly known as the "Mighty Eighth." While the fictional narrative of "Twelve O'Clock High" takes place during the earlier period, I have in most cases opted to use the more-familiar nomenclature: "U.S. Eighth Air Force."

Acknowledgements

I am grateful to the editors of the following magazines and journals, in which some of these poems first appeared—some in slightly different versions:

"**Better Hooches and Gardens**" first appeared in *Flyway: Journal of Writing & Environment*, Summer 2018, where it was also the winner of the annual "Untold Stories" contest.

"**Clausewitzian nature poem**" first appeared in *Collateral Journal* issue No. 4.2, Spring 2020

"**Contact print**" first appeared in *So It Goes: The Literary Journal of the Kurt Vonnegut Museum and Library* (KVML) No. 5, November 2016. The theme of the issue was "Indiana."

"**Humility**" first appeared in *The Line Literary Review,* Columbia University, Summer 2022

"**Kintsugi**" first appeared in *So It Goes: The Literary Journal of the Kurt Vonnegut Museum and Library* (KVML) No. 3, November 2014. The theme of the issue was "creative process."

"**Sound of a left-handed football bat, clapping**" first appeared in the chapbook *So Frag & So Bold: Short Poems, Aphorisms & Other Wartime Fun*, Middle West Press LLC, Johnston, Iowa 2021

"**10 haiku about Operation Desert Storm**" first appeared in first appeared in the anthology *Proud to Be: Writing by America's Warriors, Vol. 6*, Southeast Missouri State University, Cape Girardeau, Mo., 2017

"**Toward a poetics of lessons-learned**" first appeared in first appeared in the anthology *Proud to Be: Writing by America's Warriors, Vol. 5,*

Southeast Missouri State University, Cape Girardeau, Mo., 2016

"Toward an understanding of war and poetry, told (mostly) in aphorisms" first appeared in *The Wrath-Bearing Tree* March 2017

"a quiet professional professes through haiku" first appeared Jan. 24, 2017 with a slightly different title on Tom Ricks' "Best Defense" blog, which was then hosted on the *Foreign Policy* magazine website. The sequence of haiku was a response to Ricks' call for 150-word micro-essays regarding the question, "What should a military professional profess?"

A Few Words of Thanks

Thanks to journalist and author Thomas E. Ricks for asking tough questions—the kind that stick with you.

In January 2017, for his "Best Defense" blog at *Foreign Policy* magazine, the Pulitzer Prize-winning journalist and author solicited readers for 150-word micro-essays written on this theme: "What makes a military professional?" In the exercise, Ricks urged contributors to go beyond restating patriotic or professional truisms.

"I fear that some of our contest entries use 'professional' when they mean 'expert' or 'team member' or just 'a person of good character.' A professional should be more than that—someone who learns a body of knowledge, passes barriers to entry, is judged by peers, and meets certain agreed-upon standards of knowledge, skill and character."[1]

In answer to Ricks' prompt, I set about writing a set of haiku. My 2015 collection, *Welcome to FOB Haiku: War Poems from Inside the Wire*, included humorous haiku written about soldierly life, and I wondered if "military professionalism" might not similarly lend itself to bite-sized nuggets. My seven poems totaled only 90 words, and each delivered a warhead of wit and wisdom.

Thanks to the Iowa Army National Guard's Col. John Perkins, a.k.a. "Archer-6" and the 2010-2011 "Mayor of Bagram," who once lent me his DVD copy of "Twelve O'Clock High." This book is an echo of our many barbecue-fueled conversations about life, leadership, and how to fix things around the house and armory.

Thanks to U.S. Air Force F-16 pilot turned war poet Eric "Shmo" Chandler, author of *Hugging This Rock: Poems of Earth & Sky, Love &*

---

[1] See Ricks, Thomas E. "OK, just what does a military professional profess (Best Defense contest Round III)" March 3, 2017. *Foreign Policy* magazine on-line. foreignpolicy.com/2017/03/03/ok-just-what-does-a-military-professional-profess-best-defense-contest-round-iii/

War and *Kekekabic*, who provided his usual mix of tough love and editorial insight on an early draft on this project. Next round is on me.

Thanks to retired Canadian Army chaplain Michael Peterson for his kind and expert consideration of a later draft. Once, armed only with the solitaire tabletop war game "Target for Today" and a Twitter account, the "Mad Padre" regularly piloted epic on-line performances of immersive World War II storytelling. I hope *Twelve O'Clock Haiku* might be received by readers in a similar spirit. *"Tally ho!"*

Thanks to Pauline Shanks Kaurin, professor and Admiral James B. Stockdale Chair in Professional Military Ethics at the U.S. Naval War College, Newport, Rhode Island, for her timely insights and enthusiasm regarding this project. To both military and civilian colleagues, I heartily recommend her 2020 book *On Obedience: Contrasting Philosophies for Military, Community and Citizenry*.

Thanks to Russell Burgos, associate professor at National Defense University, whose 60-second pop-culture analysis of *Twelve O'Clock Haiku* sublimely referenced another classic film, *A Clockwork Orange* (1971). Burgos puts the cheek in national security chic.

Thanks to Brian D. Laslie, command historian at the U.S. Air Force Academy and fellow member of the Military Writers Guild, for his kind pre-publication review of this book.

Thanks to the staff at Johnston Public Library in Johnston, Iowa—particularly Dreama K. Deskins and Lori Elrick—for their assistance with Interlibrary Loan (ILL) requests and other research efforts.

Thanks also to the staff and students at Cowles Library, Drake University, Des Moines, Iowa for their assistance, and graciously extending to me special-user privileges as an alumnus. *"Go Bulldogs!"*

About the Writer

Randy Brown traveled the world as a child in an active-duty U.S. Air Force family in the 1970s, then landed permanently and happily in the American Midwest. A former editor of community and metro newspapers, as well as national trade and "how-to" consumer magazines, he is now a freelance writer and editor based in Central Iowa.

Brown embedded with his former Iowa Army National Guard unit as a civilian journalist in Afghanistan, May-June 2011. A 20-year military veteran with one overseas deployment, he subsequently authored the award-winning 2015 collection *Welcome to FOB Haiku: War Poems from Inside the Wire*. A chapbook, *So Frag & So Bold: Short Poems, Aphorisms & other Wartime Fun*, was published in 2021.

His poetry and essays have appeared widely in print and on-line, as well as anthologies. He even appeared as an "on screen" character in the 2021 *True War Stories* anthology from Z2 Comics, Denver.

Brown is a three-time poetry finalist in the Col. Darron L. Wright Memorial Writing Awards. He co-edited the 2019 Military Writers Guild anthology *Why We Write: Craft Essays on Writing War*, and curated the 2015 *Reporting for Duty: U.S. Citizen-Soldier Journalism from the Afghan Surge, 2010-2011*.

Brown was the winner of the 2018 "Untold Stories" poetry contest administered by *Flyover: Journal of Writing & the Environment*. He was the 2015 winner of the inaugural Madigan Award for humorous military-themed writing, presented by Negative Capability Press, Mobile, Alabama.

He is the current poetry editor at the literary journal *As You Were*, published twice a year by the non-profit Military Experience & the Arts. He is also a member of Military Reporters & Editors, the Military Writers Guild, and the Military Writers Society of America.

As "Charlie Sherpa," he writes about modern war poetry at: www.fobhaiku.com; and military writing at: www.aimingcircle.org.

Follow him on Twitter: @FOB_haiku

Did You Enjoy This Book?

Tell your friends and family about it! Post your thoughts via social media sites, like Facebook, Instagram, and Twitter!

You can also share a quick review on websites for other readers, such as Goodreads.com. Or offer a few of your impressions on bookseller websites, such as Amazon.com and BarnesandNoble.com!

Recommend the title to your favorite local library, poetry society or book club, museum gift store, or independent bookstore!

There is nothing more powerful in business of publishing than a shared review or recommendation from a friend.

We appreciate your support! We'll continue to look for new stories and voices to share with our readers. Keep in touch!

You can write us at:

> Middle West Press LLC
> P.O. Box 1153
> Johnston, Iowa 50131-9420

Or visit: www.middlewestpress.com